Activist Poetics

AF690215

Activist Poetics

Anarchy in the Avon Valley

John Kinsella

Edited with an introduction by Niall Lucy

LIVERPOOL UNIVERSITY PRESS

First published 2010 by
Liverpool University Press
4 Cambridge Street
Liverpool L69 7ZU
This paperback edition first published 2020.

Copyright © 2010, 2020 John Kinsella

The right of John Kinsella to be identified as the author of this work has been asserted by him in accordance with the Copyright, Designs and Patents Act 1988.

All rights reserved. No part of this book may be reproduced, stored in a retrieval system, or transmitted, in any form or by any means, electronic, mechanical, photocopying, recording, or otherwise, without the prior written permission of the publisher.

British Library Cataloguing-in-Publication data
A British Library CIP record is available

ISBN 978-1-78962-170-9

Typeset in Stone Serif and Stone Sans by R. J. Footring Ltd, Derby

Contents

Acknowledgements

John Kinsella wishes to thank his partner, Tracy Ryan, who is a part of this dialogue, and to strongly acknowledge the traditional custodians of the land he writes. He would also like to thank the editors of the following journals in which earlier versions of some of the chapters and poems first appeared: *Angelaki*, *Arena Magazine*, *Cultural Studies Review*, *Fulcrum*, *Overland*, *Poetry* (Chicago), *Poetry Review* (London), *Poetry Wales* and *The Famous Reporter*. He extends a special thanks to Fiona Sampson, editor at *Poetry Review*, for supporting the publication of a series of his ecological activist essays over a number of years.

Unless otherwise noted, all photographs are by John Kinsella.

Map of Avon Valley

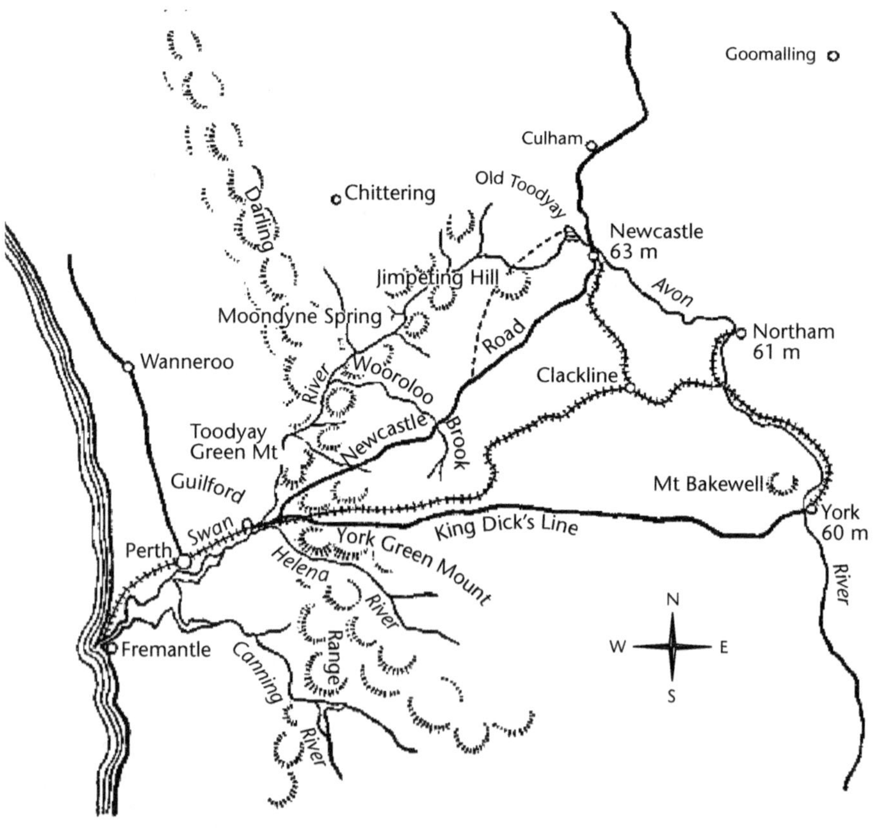

Introduction: Blood & Salt

NIALL LUCY

> But who, we?
> – Derrida, 'The Ends of Man'

Readers seeking solace are advised to turn back now. In its many eccentricities – born, I think, of a refusal to settle into a familiar style or genre – this book is not concerned to charm its way into readers' hearts, but rather to find out who might have the mettle to bear witness to it. As Kinsella puts it in the 'Preface to Paradiso' from his magnum opus to date, *Divine Comedy: Journeys Through a Regional Geography*, all he requires of readers is the resolve 'to look', the will to bear witness ... if there is such a thing. 'One doesn't have to be a believer to experience the full effects', he says. 'The willingness to look carries a certain material and spiritual recompense' (161).

Because so much of what John writes about here is of the order of 'personal' testimony, 'personal' experience and opinion, we (if there is such a thing as 'we') are asked to take it on faith. He writes, then, not as a scholar or a forensic scientist, who might be said to trade in facts that others can corroborate, but as a poet, even when he's writing prose, even when he's seemingly at his most declarative. It is not for us to 'corroborate' this testimony (how could we, since none of us is John Kinsella?), if to corroborate something is to validate it according to an ideal of disinterested knowledge. It is not for us to 'validate' what John says in these pages: how could we, since we didn't write them?

'Bearing witness, if there is such a thing', as John's friend Jacques Derrida argued, involves a leap of faith beyond 'knowledge' understood as universal and objective ('Poetics and Politics of Witnessing', 78). While this has its special moments, it is also what we do whenever we communicate with others. We don't communicate *knowing* that what we say will be received as we intended; we simply have *faith* that it will be. Without such faith, none of us – poets, scholars and forensic scientists alike – could hope to communicate at all. In this communion of 'spiritual recompense', we pledge to tell the truth and ask to be believed:

> The witness *promises* to say or manifest something to another, his addressee: a truth, a sense that was or is in some way present to him as a unique and irreplaceable witness.... With this attestation, there is no other choice but to *believe* it or *not believe* it. Verification or transformation into proof, contestation in the name of 'knowledge,' belong to a foreign space. They are heterogeneous to the moment proper to bearing witness. The experience of bearing witness as such thus presupposes the oath. (Derrida, 'Poetics and Politics of Witnessing', 82)

It's in this context that I think we're called on here as witnesses to John's own witnessing of what 'in some way' is true to him as someone who is 'unique and irreplaceable', making the question of whether we agree or disagree with him irrelevant. It's not for us to agree or disagree on the basis of 'evidence' defined as knowledge, as though it were John's purpose in writing the book that we should emulate what he says and does. The apologia (this book's nearest literary equivalent), as distinct from the manifesto, asks not that we take 'action', although of course it doesn't preclude us from doing so. It asks, rather, only for our countersignature.

What we're asked to countersign is John's witnessing of injustices that reveal themselves 'in some way' that cannot always be verified beyond dispute. His arguments for becoming vegan, for instance, may be open to conjecture, but his *faith* in veganism is no less strong or compelling for being an ethical rather than a strictly critical decision. It's this faith that calls him not only to rethink what he eats, but also to re-imagine his whole mode of being in relation to others, and for John there is no automatic privilege bestowed on 'being' as a result of being human. So while many of the pieces here bear witness to injustices suffered by Australian Aboriginal people and other marginal groups, the book's advocacy of human rights cannot be dissociated from John's affirmation of the rights of animals, plants and 'the soil'. Sometimes, indeed, his concern for the interests of the non-human verges on the misanthropic. 'Humanity', he sighs in Chapter 6, on witnessing a magnificent, four-foot long bungarra goanna get run over by a car on a remote bush road.... 'You have to try hard not to hate'.

John's been trying hard not to hate now for a long time. 'For me', he says in the 'Preface to Purgatorio' from *Divine Comedy*, 'the world is a purgatory, with hell close at hand.... One persists' (4). Often the purgatorial, if not the veritably hellish, presents itself to him as the trauma of animal abuse, even – perhaps especially – when the 'abuse' is normalised. Describing his experience of a visit to Taronga Park Zoo in Sydney, for example, which can be reached by ferry from Circular Quay and then by chair lift from the lower Harbour entrance to the main entrance on Bradley's Head Road in Mosman, he writes in Chapter 10:

> The whole business of arriving by ferry, being hauled over the circles of inferno by cable car, and being confronted by ice cream and confectionery shops set the stage. And a *stage* it was – animals in their cages and glass boxes, the absurd sideshow called 'the nocturnal house', an elephant with a genital growth, performing seals, the psychologically-wrecked apes, and so on.... I couldn't cope and didn't know why I was there – I felt lurid, like a voyeur of the most damaging kind.

Crucially, this principled opposition to what John experiences as the *truth* of animals' mistreatment, even where others may see no mistreatment at all, is grounded for him in a deeply felt relation to the land where he lives: not the land of the Australian nation state, but the land – 'the soil' – around the small towns of York and Toodyay in the Avon Valley region of Western Australia, about an hour and a half's drive inland from Perth. It's *this* land – so much of it turned into salt after years of farming abuse, so much of it stained with the blood of its traditional owners at the hands of white settlers, the military and the police – that grounds even his most universal declarations on behalf of peace, anarchy and veganism. These declarations, these expressions of faith, derive from witnessing the trau-mas of *this* land, *this* unique and irreplaceable place that calls him – no matter where he is – to declare himself the friend of the bungarra ... and therefore to be the friend of peace, anarchy and veganism. It's *this* land whose unpresentable singularity he's forever struggling to present.

This is what I take John to mean by 'activism' here: the struggle to bear witness. Because if witnessing – activism – isn't a struggle, if it isn't always *work*, then it's a prescription and not an expression of faith. To bear witness we must assume 'the oath'; to do polemics we need only assume the truth.

While John has attended his share of protest rallies in the past, his activism now is idiomatic. Bartleby-like, he prefers not to side with organised resistance while also preferring not to condemn it. He prefers not to eat meat, without condemning those who prefer otherwise. He prefers not to be complicit with a modernity that, for him, is just another name for capitalism, which could never be the bungarra's friend.

Such is the strange modality of an 'activist' poetics arising from somewhere before or beyond the opposition of the active and the passive. An imagined place, and therefore an affirmation, this 'somewhere' is also, for John Kinsella, drenched in blood and salt.

It's the place that makes his unbecoming modern, just.

CHAPTER 1

Herewith the (Auto) *Razó*: Activism and the Poet

Any book on activism and poetry is by necessity a personal one for me. 'Environmentalism', for want of a better word, is what I do in life and in my writing. However, long ago I differentiated between polemic and open-endedness, between rhetoric and, if one likes, the lyric impulse. Rarely does one write a poem of pure *anything*; but ultimately, though not exclusively, I try to keep the balance in favour of the open-ended lyric rather than propagandist rhetoric. Although I can get mighty pissed off and even write poems with subtitles like 'a poem of abuse', I try to undo my own sincerity and zeal with irony and/or figurative tugs of the carpet from beneath my own certainty. A poem in which language is not the prime generator is no poem at all for me. What's more, such a poem is a hell of a lot less effective as activism. If the reader has to work at the poem, s/he is more likely to think about the issues being explored, struggled with. Poetry should be a struggle.

I seem to be embroiled in an endless debate about the whys and wherefores of political poetry. Arguments that the poem is political regardless of intent because of how and where it is read, and the cultural implications of language selection and usage, make it a very open debate. But in essence the debate revolves more around poems as activism: the poem working for a specific political end, most often a form of change or correction. On activist discussion groups, many participants have an inclination for poetry, and poetry and 'the arts' in general become an extension of political statement. Borrowing and usage of 'found' texts and images, manipulating visual and textual representations from mainstream culture, are commonplace. [Or at least they're commonplace for a certain community of users, even while they continue to offend traditional notions of aesthetic production. See our co-edited collection, *Plagiarism! (From Work to* Détournement*)*. – NL] The familiar expression 'culture jamming' denotes an undermining of the status quo through the manipulation of 'problematic' images that have become all too readily accepted. One 'poem' on an activist discussion list that recently annoyed me and my partner Tracy involved the use of a woman with her mouth open provocatively, probably taken from a porn site, with a snappy, disposable and suitably ironic political caption beneath. In essence, we as viewers are supposed to deconstruct our own sense of viewing, to challenge the misogynistic and fetishised use of the female. In reality, we argue, one has one's cake and eats it too. The woman in the image is still exploited, and our awareness is merely ... well, lip service. This is poetry as failed activism to my mind.

A recent discussion with an anarchist activist friend of mine involved differentiating our respective anarchisms. He termed his 'aesthetic', and mine 'moral'. It's an interesting distinction, though no differentiations are decisive. When I write an 'activist poem', it *is* for a moral purpose, but the reason I select poetry as a form of expression is as much because of the constraints as of the freedoms of the 'genre' (I don't like categories!). But then, the most polemical poems (and thus the most ineffective outside their immediate moment of action, I'd argue) often use absorbable sing-song and banal rhythms and strict rhyming patterns: these are 'aids to memory', mnemonic devices to allow the message to be carried by word of mouth. Oral poetry, sure, but oral poetry resisting the change that comes with slippage, error, and interpolation – I am talking here of the polemical poem that means exactly what it says, serves a specific purpose, and is never intended to grow with the telling. By contrast, when I write an 'activist' poem, I want it to be misinterpreted as much as interpreted; I want even something that seems overt to be questionable – at least through that basic and fallible device: irony.

One of the objections to the activist poem is that its range is necessarily narrow. Motivated by an extreme of emotion, it doesn't allow readers themselves to invest a personal scale or chromatics of emotion. I challenge this: an activist poem should motivate investigation, and not tell readers what to think. Data are data, as subjective as the presentation might be. Data can find their way into the poem, but the figurative medium mediates our response to 'fact'. Fact can, in fact, become misdirection, and certainly misprision.

While many might argue that prose would be a more effective tool for the activist – being more to the point, and more direct – I would argue differently. However, prose might form an important prelude, even component, of poetry. Be it Dante's *La Vita Nuova*, or the *razós* written to accompany troubadour songs, explication is an active and integral part of poetry. Whether in the glosses of 'The Rime of the Ancient Mariner', or the prose 'responses' within the haibun, prose becomes a counterpoint. This prose doesn't have to 'explain' a poem – in fact, who really wants a poem explained? – but it can interact and hence enrich, and highlight or offset the poem. A bit like a frame around a painting, or the space around a sculpture. This essay, then, is an (auto) *razó* for an activist poetics.

This essay is also really a long *razó* for a life of writing poetry. In the superb volume *Razos and Troubadour Songs*, translated by William E. Burgwinkle, the introduction opens with a paragraph useful enough for my purposes here to quote in full:

> The usage of the Occitan word *razo* within the body of thirteenth- and fourteenth-century texts collected in this volume reveals how fully medieval authors came to exploit the semantic richness of the word's Latin model, *ratione(m)*. The most generalized and reductive definition, and the one which came in time to define a literary genre, is this: introduction, explanation, reasoned commentary, subject matter, background, gloss. Indeed, we do find that all the short prose texts collected here clearly serve in one way or another an introductory function in relation to the songs with which they are paired. Referring to, and citing, specific lyric texts composed by troubadours of Southern France, Northern Italy and Spain, the author claims to provide an explanation for why a particular poet composed a particular song or songs at the moment that he or she did. At the same time, he does not treat the song in a vacuum; rather, by adopting a uniform rhetorical practice and vocabulary, he relates the song to a larger social and literary phenomenon: the composition and exchange of vernacular songs of praise in the Southern French courts. (xvii)

Before I take a public action – write a letter of complaint, give a public lecture, attend a protest – I inevitably write a poem on the subject. Poems usually follow the action too. And might continue to do so long after the event or issue has passed, even if there's been a negative outcome (a forest block has been logged/mined/cleared, for example). The writing of poems becomes part of a mantra of witness and empowerment. Often, poems form a visceral and literal part of a protest. In what follows I am going to explore two recent activist issues, of very different but still related natures. I will then chart their appearance and digressions in poetry. Basically, the issues are as follows:

- visiting a forest protest site in the Arcadia jarrah forest in the south-west of Western Australia to lend support to activists
- involvement in a recent *Indymedia* discussion/protest against the use of violence by activists at the G20 conference held recently (November 2006) in Melbourne.

Each of these 'actions' involved the writing of prose and poetry.

The Net, especially the *Indymedia* site, became my source of information regarding logging activities in Arcadia Forest. This regrowth jarrah block located near the Wellington Dam catchment, not far from Collie (a coal mining town), is contentious for a number of reasons. First, it's native forest; secondly, the regrowth since the 1930s has been considerable, and it is a substantial forest of jarrah and marri, with thick undergrowth (snottygobble, etc.); and finally, it forms an integral part of the extremely rare and critically endangered habitat of the marsupial land quokka (the quokka is also found on Rottnest Island off the coast of Perth, the island's name having been given to it by seventeenth-century Dutch sailor Willem de Vlamingh, who mistook the quokkas for large rodents and accordingly christened the island 'Rattenest', meaning 'rats' nest').

Together with an old anarchist friend of mine, I visited the protest camp that had been moved on by police from its original site within the logging area. The preceding days had seen a lot of 'locking on' activities: the use of 'dragons', concrete and metal anchors, in essence, with arms literally concreted and welded into large objects so the police and loggers can't move protesters without the risk of injury to those protesters (they are usually defeated, however, with an array of cutting equipment); climbing trees and roping oneself to trunks or branches; and lying in front of bulldozers, chaining oneself to machines, etc. These direct actions had had little effect. The best such protesters can hope for is intense media interest, and maybe delaying tactics, with the possibility that other forms of protest will prove effective before further damage is done.

Visiting the Arcadia Forest protest campsite convinced me that a holistic plan of action is the only way, a plan to beat the loggers at their own game. I have attended many such protests, and have grown increasingly frustrated with these activities. The government and the companies involved always contradict their own rules: entering dieback areas (dieback is *Phytophthora cinnamomi*, a fungus/pathogen that attacks the roots of trees – it is spread by movement of people and vehicles through the forest), making use of streams, crossing boundaries. Thus I have argued that a legal, written and photographed approach, to highlight the rule breaking, is essential. Recent legal challenges (by Federal Senator Dr Bob Brown in Tasmania) on behalf of protected flora and fauna have been effective, and I have been told by environmental lawyers that this approach is viable. So I sent this letter to protest organisers:

> Hi...
>
> I think what we need to do is prepare some kind of doc of aims in point form. e.g.
>
> 1. To preserve native forests – old growth and previously logged forests. To preserve and protect the wildlife and vegetation therein.
> 2. To minimise disruptions to native forests. To encourage alternative ways of viewing the intense uniqueness and splendour of these places.
> 3. To encourage the planting of trees on already (!) cleared land for use as 'timber products'.
> 4. To establish a fund to be used to defend native forests in court – i.e. through legal avenues.

5. This fund to be used to allow preservation activities to be pursued without the fear that government and private companies can act with impunity because they know those opposed to these activities have no resources to go to court. Very often, loggers and miners and pastoralists act against the law itself (spreading dieback, polluting water, going outside agreed upon boundaries), and no one can afford to challenge them. There are innumerable examples of this and the perpetrators should be held accountable (under their own – inadequate – rules!). Under various wildlife and environmental legislations, there are strong legal grounds for the cessation of logging and usage of native forests.
6. To establish a pool of interested legal people who are prepared to work for little or nothing.
7. To establish a charity or related status (i.e. creating a position/structure re dealing with donated funds and using them for legal purposes).
8. To actively promote such a fund to interested parties. A clear outline of how the money will be used and how it will be managed is essential and the key to receiving donations.
9. Those we (the group) take legal action against should know that we are financially capable of losing. That's the trick – not that we only hope (expect) to win, but that we can lose and then have enough up our sleeve to appeal and to mount another legal challenge in a different area.

This is just off the top of my head – interested people should feed into such a doc – consensus is essential. Once we have these sorted, I think we should look for donors – and I am very happy to work towards this end.

best,
jk

Now it might seem there is little to connect this with poetry but, in fact, there's a poetics in this approach. As an anarchist, interaction with legals in this context does not appeal to me, but in the end they are the translators of language into fact in the structures with which one has to deal regarding this matter. They take the kernel of the poem and write the *razós* that come before it. As a connected action, I took the poems I had written about this issue, and circulated them. I had done the same thing after attending the Ludlow Tuart Forest protest near Busselton in south-west WA a few years before.

Poetry is an effective way of taking the discussion out of 'interest areas' into a broader discussion. I would argue that the metonymic connection between the trees we exploit for paper, and our writing, is just cause for investigation of source and culpability on behalf of writers. I gave a workshop a few years ago when two smart teenage girls spent their time mocking environmentalism in writing for this very reason. It's an obvious target, nothing really smart there, but I had to pay them credit for their consistency. The issue, the contradictions, weren't going to go away. Computers excepted (and they too bring their own myriad of environmental issues), paper is the tissue of the writer's body. I was down south last week [going 'down south' is a Perth expression, with 'south' referring to anywhere south of Bunbury, at a pinch, but more commonly to the vast rural and forested areas south of Busselton and stretching inland to the South Australian border – NL], checking out plantation timber: native bushland surreptitiously cleared, the use of pesticides and herbicides – the golden dream is not always so golden. It needs vigilance, monitoring, and writing about. What we write with and on are part of the responsibility.

The problems of the plantation timber industry aside, the starting point is the preservation of native forests. Here is a poem written in protest at the logging of jarrah in native forests in Western Australia. It was written prior to the Arcadia protest, but with the same region and actions in mind. The first two lines are taken from 'Love Sonnet' XXXI by Zora Cross, an Australian poet who wrote largely during the First World War and who's best known perhaps for the 60 love sonnets in her 1917 book *Songs of Love and Life* [see John's *Love Sonnets* for his sequence of 60 poems that reproduce, in numerical order, the first two lines of Cross's sequence; note too that John has republished several of Cross's poems over the years (see for instance his edited collection *Dreamhoard*) – NL]:

XXXI

We must look around upon our children dear,
Living through them, this present and that past;
And this is never more evident
Than when walking through old-growth forest;
The toddler ambling ecstatically the filtered light,
Rough-barked jarrah towering and closely packed,
Birds endemic to that place only
Darting around the undergrowth; his pleasure
Palpable, loggers nowhere nearby,
Though yellow markings on trunks,
Pink ribbon fluttering down the valley,
A 'grammar' of *his* children's barren landscape;
Denecourt's wild walks take us nowhere here,
A lust for aluminium undoing the propaganda.

There are connections between the naturing of the poem and the naturing of de-natured lives: the European wild walk as entertainment and reconnection with lost primeval roots is played with in relation to the colonisation of 'Australia'. This is a poem that celebrates place, and laments loss, but also tackles language as the generator that makes such colonisation possible and likely, as well as making it possible to challenge it. Language and activism co-exist tautologically, capable of achieving the same ends, and also of being mere distractions. Activism as an act of mere habit is like uttering clichés without respecting and experiencing the richness of the clichés – their necessity for existence, their different inflections and implications with every use.

I chose this poem, in the context of this essay, for the last line: the jarrah is also cleared for mining bauxite, which is then turned into aluminium. Most of us make use of aluminium, including the locked-on activists: in their dragons, and in cooking their meals around the campfire. The inescapable irony in action becomes the undoing of the poem in language. 'Human kind/Cannot bear very much reality'. How many of my essays have quoted this truism from T.S. Eliot?

The following poem was written sometime after visiting the Arcadia activist site, but with the future activism outlined above in mind. It is part of my version of Dante's *Paradiso*, in progress at the moment [since published as one of the canticles in John's *Divine Comedy: Journeys Through a Regional Geography* – NL]. The allusions are all to Dante's twenty-first canto: the contemplative, the golden ladder, predestination, and the usual condemnation of papal corruption. The ingredients are all there for a piece of activist verse that is also a nature lyric (after all, that's what I basically write), and a piece of metatext:

Canto of Arcadia (Saturn, 21)

She won't smile the smile of incineration.
They won't sing the song of disintegration.
Together sprung, noise music, see-through mirrors.

Locusts spin in flurries. They have begun to fly.
The bobtail out the back is fat. Crow feathers
are strewn over acres. Rings forming

in the bright sky. Their descent is known.
The ladder is propped up against the rainwater
tank – I have to ascend to clear the leaves

from the grille, but she would only have me descend:
precarious, I perch. My ears are giving me hell.
Two days ago we drove down to Arcadia.

Into Arcadia along forestry roads, rutted
by logging trucks filling the quota. The protest
camp with its wire and papier-mâché

quokka. Mainland quokkas ranging the jarrah forest.
Four hundred metre corridor – promised – barely
two hundred. The sacred stream, arterial, filled

and forded. In and out of dieback like humour.
Fifty-year regrowth cradling the undergrowth.
Marri trees less useful to Gunns – horror

company – left as Habitat. Marked H in rings.
Occult emblems glowing out of brown-greens.
Ensconced in the forest, fallacy is sucked

into machinery, police wagons. About the campfire
dreadlocked souls discuss lock-on methods. Concrete
and car bodies and piping and their own impact

on the forest. I am here now, the soul glow still
intense but tremulous. Camp dogs snap at flies –
that's memory, and taking the message out

is what's supposed to happen. To be arrested
is to arrest attention? To believe in something
is better than believing in nothing? Tree by tree.

She won't smile the smile of incineration.
They won't sing the song of disintegration.
Together sprung, noise music, see-through mirrors.

What makes the poem activist is that it leads a dual life as a piece of poetry (intended to be in a book), and a piece used to provoke discussion and debate over a specific issue. It fits in neither place comfortably; neither do I want it to. It is a hybrid.

Moving on to the second broader issue I outlined earlier – the *Indymedia* discussion (at Perth and Melbourne sites) about violent action at the G20 summit – 'the poem itself' becomes an integral part of an activist response to an event. After the physicality of some activists at the protest (which I did not personally attend), 'response' is everything.

In a nutshell, a group of masked and hooded activists calling themselves 'Uncitizens of Arterial Bloc' committed acts of violence against police vehicles and barriers, and then defended their actions in a manifesto posted to *Indymedia* and various activist discussion groups on the Net. Their claims of being every person, of defending the purity of their anti-capitalist motivation, and that their acts of violence were against property and not people, didn't wash with me. This is my response to their communiqué that defended their anonymity and violence:

> A disguise is a disguise is a disguise
> by John Kinsella
> 2006-11-23 8:04 PM +0800
>
> Well, as an anarchist who is happy to be identified both positively and negatively, to be othered and included, to be mocked and appreciated, I find this laughable. Actually, I think it might well be the Ern Malley Hoax [the reference is to the fictitious 'Ern Malley',

whose poems, written in an afternoon by James McAuley and Harold Stewart, were published by editor Max Harris in a special issue of *Angry Penguins* magazine in 1944 as the work of a 'lost' modernist master; Harris was turned into a figure of national ridicule when it became front-page news that McAuley and Stewart had duped him – NL] of protest/activist (and I do differentiate between the two here) 'politics'.

Disguises do not uncitizen one, but rather make one complicit by subverting identity to what the State wants – conformity of appearance. Suited up in your whites or blues, you do no more than performatively ironise the position you occupy – and that's fine, but let's call it performance and be done with it. As a vegan pacifist, I find it doubly ironic that any conversation such as the one above (conversation? validation?) can be had without mentioning the massive state and INDIVIDUAL CROSS-CULTURAL intra-spiritual etc. etc. abuse and use of animals in most projections of power.

In your suited diversity, taking on whomsoever, how many of you snack on a burger or a piece of cheese or the numerous e-numbers that constitute the backgrounds to even seemingly sound 'vegetarian' food. Or are we separating oppressions? The answer according to PETA [People for the Ethical Treatment of Animals] is to get nude – to defamiliarise (but in fact the opposite) through the sameness of the human body. Or maybe to pour red paint over furs (how much thought is put into the origins of that paint? – I mean the REAL origins – the nature of the labour and systems that allowed for its production and sale ... or how the activists got to the scene of the action?). I digress?

Semiotically, the above 'communiqué' plays every second-grade post-grad trip and trick I've come across – and in doing so, establishes a hierarchy of knowledge, terminology, and access. The wonderful thing about disguise, as any carnivale player will tell you, is that it is tantalising in its prospects of revelation, non-accountability ... the list is endless. The serendipity and occasionalism of the action – the turning-out through costume of oneness – suggests a consensus of viewpoints that I believe is duplicitous. Costume stands for something, as does anonymity (come on ... how anonymous to friends (etc.) are we?). And of course we don't want to be – on our own it's frustrating not to have someone to alleviate and placate our narcissisms ... for that's what they are.

In the horrors of a language of unified State oppression that surrounds us – with its attendant bigotries and lies – all we have to give are our identities. The numbers they have for you, the details for almost all of us – certainly in Australia – give us something to throw back at them. Devaluing their values comes through showing we don't respect them, do not recognise their claims to authority. Stop making use of the tools of the State. We ought to leave the electronic world (yes, hard to get to from the street, ain't it!), banks, public transport, air travel, movies etc. etc. We all should – it is my aim. Don't use and abuse animals. I've stopped doing that (as have many others) for twenty-three years and am

> happy to stand up and say, hey, this is what I've done. Just don't participate – it's the most powerful resistance of all. Then when you stand up against wankers at G20s and elsewhere, it's as a genuine resister and not as an appeasement of the vanity and absolute sense of inadequacy we all obviously feel. Cutting your credit cards, or social security details, or any of the other bits and pieces we all have that connect us to the State in full sight of the police and ASIO [Australian Security Intelligence Organisation] cameras would be far more effective. They'd say – well, it only hurts the protesters. But hey – it hurts them more.
>
> Oh – and excuses for violence are laughable.

There were many responses to this post. Other participants criticised me because they believed I was advocating the need to 'opt out' of community, whereas I meant only to opt out of the machine of the State. I responded as follows:

> Non-violence is active resistance
> by John Kinsella
> 2006-11-24 7:58 AM +0800
>
> 'Opt out' is one of those 'nimby'-like [as in the British colloquialism 'not in my backyard', and not a reference to Nimbin, a small town in northern New South Wales where it's de rigueur to be 'alternative' – NL] expressions – it's designed to marginalise before conversation begins. I believe the State should be challenged at every opportunity, and that oppressions should be resisted – non-violently. Violent acts seem not to have yielded much positive outcome to me – NO progress has been made anywhere (at least that has lasted beyond the occasion. Check out the aggressions of men at Greenham Common when they were asked to leave the group/s. One can resist violence – the fighter planes and bombers – and yet be enacting it against the people one lives with … doesn't add up to me). As long as people can be violent – directly or indirectly – to animals, for example, they are operating in a world of violence – so, sadly, defences of such 'actions' are to be expected.
>
> I didn't say non-participation alone is the answer, but certainly participation is a major factor. The sheer evidence of this computer and the implications of internet usage makes hypocrisy of my interaction here unresolvable – I should, and will, abandon this technology. As long as one is connected to the power grid provided by the State, one is culpable.
>
> Elsewhere someone mentioned the construct of identity – yes, it is that, and that's why we can throw it in their faces. It's what we have that they recognise as 'useful' that we can undermine them with. When people say 'some of us are already doing this' – cutting off from using the State's facilities etc. – I ask, how much? The fact of this interaction suggests otherwise.
>
> Back to my earlier point/s. 'Technology' seems the trap here. Violence not directed at living things is still violence – smash a

police van and another is manufactured. The tools and materials for making those police vans come from the same source as the manufacture of the cars and buses and trains we all ('here', at least), make use of. Thus we are complicit.

I believe in the empowerment of direct action non-violent protest. I think that one should stand in front of the bulldozers (yes, and one might be bulldozed – the act of allowing it to happen does engender some degree of violence, but here I differentiate – a passive-positive action, the use of one's body to absorb the violence of the State/business/violent parties, seems a reasonable self-choice to make), but not wreck them (though I think they shouldn't exist in the first place – that is, manufacturing of such machines should cease).

I believe one should gather in protest against G20s – not turn the act into a self-serving carnivale that celebrates nothing other than one's own frustration and need for social subservience to one's peer group. Action is driven by necessity – and that's all there is in the wrongness of the world we live in – necessity for action. To me, the sort of action we take matters. Disabling one F-16 does not stop the state-corporate machine that makes the F-16. The State dissolves when we don't do what it tells us to do. It might well kill me for not doing what it wants me to do, but I certainly won't give it the excuse of violence to achieve this end result for its propaganda outlets.

Someone mentioned (somewhere) the Angry Penguins and 'smash the state' in one sentence – what a joke. The Angry Penguins did nothing of the kind – they were largely 'middle-class' self-serving artistes who spent most of their time defending themselves from attacks from the left whilst pointing out that they published 'real' Marxists. Take a look at the copies of the Angry Penguins' magazine – lots of pseudo revolution and very little substance outside cultural discussion.

Nothing can be said without implications – nor written. Context might be vital, but words create their own context as well. Quoting Orwell in any way, for example, undoes itself by the fact that the man was a spy against the left he supposedly supported. Oh, and as a vegan, every day is part of a direct action against the violence of animal abuse – what you choose to eat is at the core of who you are, I think. And who WE are. Self-righteous? Everyone with a strongly held set of opinions is. That's what helps us resist. Every small action we take matters – and that's why the how and why of those actions that take place matter so much. I am not saying that one has to be a vegan to make an act valid – I am simply trying to point out that every action has an implication we often don't see or expect. The decision I made many years ago to renounce violence (against everything) has empowered me to be held accountable in ways I would never have owned up to. I was once a very violent person and it saddens me to have to recognise this. Oh, and getting beaten up by the forces of the State doesn't stop them doing it to the next person. I have seen horrendous

> things happen in lock-ups in the '80s, and, as a consequence, I feel a need to be as effective as I can in my resistances. Non-violence – apart from being an ontological decision – places me in a position to challenge the State and their corporate buddies far more effectively. Self-righteousness? Probably – but I believe it can work. So, opt out, yes! And participate in refusal. Both.

As can be seen, poetics were never far from my response. In the ways that matter to me, the above is really an *ars poetica,* sans a discussion of lyricism! I cannot quote the responses back, but some were quite ferocious. In essence, their defence of violence was paramount. In the end, we get trapped in our own righteousness regarding an issue, and poetry is the only possible mediating force. It usually brings three reactions: empathy, hostility, or 'I don't get it'/indifference. Most fascinating, however, is that even though one clearly has an agenda in posting it, especially if one has taken an overt political position beforehand, the ambiguities of prosody and the figurative bring a bizarre calm to the argument. Now, a straightforward rant or polemical piece would likely achieve the opposite, but the poem as ambiguous mediator, as conduit for a 'new language' of discussion allows movement:

Graphology 639: The violent disguisers

Traumatised by the 'psycho violence'
of NVDAs –
  they'd crash a truck
through police lines
    and grow giddy
  with the spilling
of images.

Comrades, they'd say, Comrades,
it's just a different
tactic,
  we're all on the same sade (I meant to write 'side'
but am reading Freud's *The Psychopathology of Everyday Life*
so will let the slip stay...)
                        of disguise, communiqué,
  uncitizening:

hooded (not veiled)
they'd hook a catchcry,
let the wealthy hatch conspiracies,
for they too get hungry
and have their disabilities.

I was asked recently by the editor of *Stylus Poetry* magazine: 'Does writing poetry influence your other writing in regard to approach or technique?' I answered:

> I have always been interested in the blurring of genres. I attempt to operate in a liminal zone, and think that it's pretty hard to leave this liminal zone even if one wishes. There are no categories – only flow and hesitation.

And that seems to be what happens when the non-propagandist (I like to think – though this is all relative and subjective) activist poem – not a contradiction! – is deployed by way of 'answer' and response.

In *The Ecology of Freedom*, Murray Bookchin laments our contemporary inability 'to sense the wealth of subjectivity inherent in ourselves' (279). The poem morphs the argument because it connects physicality with mental activity, flesh with emotion. A poem is an extension of the body: of self, of community, of body politic. When I write an 'activist' poem, it is not a matter of getting on a soapbox, but of placing my body in the line of discussion.

I heard a wonderful segment on national radio this morning about a nudist (naturist) colony open day in Canberra. The spokesman said that after a while, everybody there basically forgot they (and others) were nude. I wondered, then what's the point? But there is a point: one can enjoy the 'freedom' of the body, and the enclosures of social discourse. For social discourse is never free, and the manners that mediate our interactions are just a degree removed. Unclothed, sure, but to stare at each other's bared genitals would just be downright rude and inconsiderate. Thus the poem becomes a set of manners, an acceptable (if sometimes difficult to comprehend) code of conduct while laying 'oneself' bare. As Bookchin puts it:

> In summary, human rationality must be seen as a form and a derivative of a broader 'mentality', or subjectivity, that inheres in nature as a whole – specifically, in the long development of increasingly complex forms of substance over the course of natural history. We must be very clear about what this means. Natural history includes a history of mind as well as of physical structures – a history of mind that develops from the seemingly 'passive' interactivity of the inorganic to the highly active cerebral processes of human intellect and volition. This history of what we call 'mind' is cumulatively present not only in the human mind but also in our bodies as a whole, which largely recapitulate the expansive development of life-forms at various neurophysical levels of evolution. (279)

This is a *razó* to a would-be troubadour's song, if ever there were. Critics have applied 'Bookchinian' readings to my work in the past, and mostly I think these are off the mark, but by allowing more metaphor in his prose than Bookchin (I think) intended, and somewhat decontextualising, the point about interactivity seems to me spot on as a vehicle of poetic process. One of the problems we encounter when objecting to a linking of poetry and activism, especially in locating the prescriptive 'occasionality' of a poem written because one feels the need to say something, or responding out of emotional 'excess', is that it lacks inspiration or linguistic causality.

For many poet-activists for whom writing is a political action – and I would include poets as diverse in their poetics as 'Homer', Shelley, Anjela Duval, Judith Wright, Andrea Brady, Lionel Fogarty, Ouyang Yu, Benjamin Zephaniah, Janet McAdams and others in this picture – there is a fluidity between the 'passive' and the 'active', and interactivity of body and mind. Interestingly, the body of the activist who risks life to chain him- or herself to a bulldozer reverses the equation somewhat, though this is not to deny the 'found' poetry of the moment. Performance is also poetry, and the actions and sounds of struggle emit their own prosody.

Back in my 'bad old days', when I got myself in a difficult physical or mental state, I would recite physics formulas like a mantra, to convince myself I was still alive, and to keep myself calm and in check. V squared = U squared + 2AS; V = U + AT; S = UT + half AT squared, etc. Tongue twisters, childish rhymes, whole Yeats poems, all do/did the same trick. In the moment of activism, poetry becomes a binding force. What Bookchin didn't add, of course, is the interstice with spirit/spirituality. But that's a personal angle....

If the famous Bertran de Born's song is often overtly political, it's the less politically referenced troubadour songs that interest me as activist. In the *razó* to Guiraut de Borneill in the Burgwinkle translation, the *razó* writer says of Guiraut: 'And while he was there he dreamt a dream, which you will hear about in this song, which says:

> I cannot prevent that to the pain...' (23)

and then the song of the troubadour follows this prose explication, only enriched for the introduction, the context, the curating. The troubadour finishes his song with the lines:

> And toward this end I have made a real effort
> That you might understand every song that I compose. (25)

He does so in order that 'his' or simply 'the' pain of (possibly unrequited) love and the fact of honour and patronage be clear, much like the activist's intent.

There's not much of a gap between love and activism and the creation of a poem. Apart from protesting against a destruction, the activist poem also attempts to provide moments of the beauty/'good' one is attempting to preserve. Even out of loss, a future positive might be achieved. And this is how the song begins – and I concur with its 'issue', with the association of pain and beauty:

I cannot prevent my tongue
From moving to the pain in my tooth
Or my heart to the new flower
When I see the branches blossom.
The songs of birds in love
Are heard through the woods
And though I am troubled
And look unhappy,
When I see songs and gardens and fields
I am refreshed and take some comfort. (23)

CHAPTER 2

Standing Up to Aggressors

A Template of the Poet as Activist

For me, poetry has no point unless it's a prompt or aid to political and ethical change. This is not to say that a poem should be political or ethical *instruction*, but rather that it might engender a dialogue between the poem itself and the reader or listener, between itself and other poems and texts, and between all of these and a broader public (whatever that might be). I see myself as a poet activist – every time I write a poem, it is an act of resistance to the State, the myriad hierarchies of control, and the human urge to conquer our natural surroundings.

In using the language tools I have inherited (and, at times, seek to alter neologistically, even to dismantle), I am inevitably part of what I critique. I am complicit. I try to lessen the ironies of my own life in terms of what I say (by being vegan, giving up flying, resisting the logging of forests and so on), but I am still participating in social discourse and 'making a living', and that paints me into the same corner as most of us. The very act of using a computer contradicts the impetus of de-technologising that underpins what I do. Furthermore, I accept that what I see as core ethical issues are not perceived that way by many readers of poetry.

I should say that, ironically, I come out of a 'background' in science (in terms of my early interests), and that at an early age I was working in labs and obsessed with the nomenclatures of science. In subscribing to what I term a 'neo-Luddism', I am not opposing the accretion of knowledge, but the misuse of that knowledge. So much 'science' is a desire to control and to profit. This I oppose. The language of poetry, even in its most lyrical modes, is a language of specific usage – 'poetry' is about arrangement, selection, and presentation as much as an inherent value in what's said. That process of knowledge regarding expression is, to me, scientific.

Science per se is a process of investigation/observation/patterning through to hypothesis, and rests in that which can be scrutinised and in systematic acquisition of knowledge. That's what a poem is to me too. This is why an activist poetics doesn't have to be subjective propaganda – the subjective has a part in it; indeed some would say that without this a text can't be a poem – but if a poem doesn't utilise knowledge and the *processes* of obtaining knowledge, then it does less work than it might towards resisting damage (to people, to animals, to plants, to the land itself). I am interested in extensive digression, degrees of separation, even verbal tricks and diversions – in other words, a circuitous route to discussing or seeking to discover a 'truth'.

No poem really *knows* a 'truth', but it has knowledge and offers ways of approaching truth. The use of language is precise, even when it gives a semblance of the unconscious, even when it is automatic writing. In the surrealist sense, the conducting of automatic writing exercises was experimental textually and scientifically, and was as much about the act of recording the data of process as it was of the subject connecting with the unconscious. It was, at least, quasi-scientific.

And the pseudo and the quasi interest me. The games of dismantling and rearranging, of exquisite corpse and chance, are all part of the science of a poem for me: they are just different systems of knowledge. That's why an activist poetics can include the radically linguistically innovative, as well as the straight declaration ('logging that Tuart forest is wrong'). Parataxis, conventional end-stopped lines and enjambment, narrative description, metaphor and metonymies are all part of a process towards confronting hierarchies and imposed structures. We work from inside to open a view of the outside, but not one that destroys in the process.

A pacifist, which is what I am, can be the strongest resister, and pacifism the most defiant form of resistance. Same with language usage: I mix the old and the new to engage with a debate about protection, preservation, conservation, and respect of the 'natural' world. I am aware of the problems these words carry in terms of implying complicity, because I am a poet rather than a speech writer. For me, because of this, poems can stop bulldozers. Not because they *just* say 'stop bulldozer', but because the intricacies of language challenge, distract, and entangle the bulldozer. I am using a semantics not of analogy, but of opposition. My words are intended to halt the damage – to see what shouldn't be seen, to declare and challenge it.

In Warren F. Motte Jr's remarkable book *Questioning Edmond Jabès*, we read of an insistence on linguistic hierarchies (down to the 'letter' – 'an authoritative integer' – and what Jabès terms, as Motte notes, 'the grandmother of the word') that I recognise and acknowledge personally, but also seek to subvert in very specific, personally motivated 'activist' ways:

> Other, more venerable traditions are also reflected in this aspect of Jabès's poetics, most strikingly perhaps the theories of a thirteenth-century cabalist named Abraham Abulafia, who elaborated a

> system called *Hokmath ha-Tseruf,* or 'the science of the combination of letters.' Therein, letters and their configurations are used to achieve a mystical state of meditation; Abulafia likened it to 'a music of pure thought,' the alphabet supplanting the musical scale. As Abulafia's combinatorics result in contemplation of the divine, so Jabès's result in the Book, that highly personalised construct whose multiple attributes approximate the divine. (33)

While Jabès's poetics of witness and exploration of heritage and learning (which I am not attempting to apply, or in any way appropriate) differs from my own poetry of resistance and poetics of activism, I will say that, in terms of the damage being done to both the human and non-human environment, the deployment of language in such a context necessitates a scientific system of language use that also contradicts the terms of its own production. The divine for me exists in existing – I require no spiritual certainty to know that preservation of all life (animal, human, and plant) is the desired outcome of my poetry, and that the gaining of knowledge is concurrent with that process.

Over the last thirty years of writing poetry, I have utilised a process I term 'templating' to create a dialogic structure between a foundational or templated work and the one I am writing which converses with it. From tracts by explorers, through to Philip Sidney's *Old Arcadia* and Dante's *Divine Comedy,* I take a pre-existing 'historical' and cultural artefact – a text – and write away from it. Doing this with 'classics' of European literature has many implications when set in a colonised space such as Australia. The act of 'transplanting' as part of a process of 'de-colonisation' of language becomes a dynamic, activist, and anarchist process for me.

I have written about this process elsewhere (see *Disclosed Poetics* and *Contrary Rhetoric*) so shall only refer to it briefly here. What it allows me to do is critique my own processes of cultural inheritance and usage, and thereby place the issue I'm exploring in that 'critiqued' light before I begin. Environmental damage was still environmental damage in Dante's time as much as now, but its social implications in terms of human values and society might (might!) have shifted in some places and at certain points of cultural reception. I use those points to define my activism. For those interested, I have articulated these issues in the introductions to each canticle of my 'distraction' of the *Divine Comedy* – it's not a translation, but an independent set of my poems working through and referencing the original.

The work I have recently started on (see Chapter 11 and Coda) uses Henry David Thoreau's *Walden* in a similar templating way. It's a framework I distract and reflect upon to make points about ecological issues now. I don't write 'over'; I write away from the original as I have done with my previous variations, distractions, and departures from other works. I look to create an alternative science of 'the combination of letters' to undo those sciences, as much as discourses of pleasure and leisure that profit from, ignore, and disrespect natural environments.

Knowledge has been abused in so many ways and it needs reclaiming as a means of protection and preservation, rather than as rapacious exploitation and a means of increasing (some!) humans' comfort and satisfactions (whether they call these spiritual or material). Having lived in America (our son is 'American' – though I am no believer in nation wherever it lays its claim), England, and Australia (where I am from, and the place I mostly write about), *Walden* allows me a full canvas to explore local and international issues of place, belonging, and environment, and the cultural politics such an investigation conveys. I affirm Indigenous land rights, I affirm local ecologies, and I affirm a science of language as an act of resistance.

Activist Poetry 2: Foxes and Corellas

Driving down to the city this morning, we saw five or six emus crossing the road in an area of national park where I hadn't seen emus before – not once in a lifetime of travelling that way. It was a remarkable and invigorating sight as they plunged into the wandoo woodlands, negotiating their way through the spiky hakeas and parrot bush.

On a personal level, it came as a kind of foil for the weekend-that-was – a complex amalgamation of environmental affirmation and also witnessing horrific environmental crime. The sort of weekend that leaves you wondering if any form of environmental activism has any chance of succeeding, yet nonetheless also convinced that there is no choice about acts of resistance. Without them, the environment has no chance.

And writing a statement such as that is part of a process of creating poems that hopefully resonate in different ways and in different contexts, and extend what is a particularly local debate into the wider dialogue of which, sadly, it is also part. The compulsion to witness in poetry, the desire to overcome a feeling of crushing failure, and the need to create a cautionary tale that is more than propaganda, that entails its own neglect or active refusal by recipients of its message, goes hand in hand with a volatility and (maybe overly) emotive response to the situations as they happen.

I can see the poem forming in my head as I am raging against an act of destruction, though not as a fetishised aesthetic 'response', but in the struggle to formulate a language of reply that is not aggressive and thus self-defeating and hypocritical. I am being somewhat obtuse here. To begin at one possible beginning....

On Friday night at about 9, Tracy asked me if I'd heard a rifle shot. I hadn't, but I did hear the one that came just after she spoke. A few minutes later: another, and much closer. I went outside and wandered into the dark of the block, to detect torch lights up the bushed laneway, next to the small road reserve. Another shot – maybe two hundred metres away at most. It was the report of a high-powered rifle.

I yelled at the top of my voice to stop, the torches swung in my direction, and there was another gunshot. Inside, Tracy thought I had been hit; for a moment, so did I. Eventually, though we're a long way out of town, it took the police to resolve this situation. Never a desired intervention, but it has to be said that if they hadn't turned up, one of us mightn't be alive.

It turned out (we found on Sunday afternoon) that there was a fox hunt being conducted in the area. Fascinating, how private land which people around our way defend with such passion should change into public land without boundaries when pursuing foxes – the great hunter-capitalist liminality! – and that reserve land, where shooting is illegal, should become part of the script. That's part one of the shire-as-killing-zone.

Now, 'part two'.... Next morning. Saturday morning about 8.30. A cascade of explosions in the distance. Sounds like firecrackers or a fireworks display. It can't be – no events on, and the wrong time of year ... the whole district would erupt in flame as everything is still tinder-dry and there's a full fire ban.

The sound diminishes, and one can detect the tone of individual shots being fired – shotgun shells. I decipher that it's happening in town, probably by the river. Corellas. They are culling corellas. Someone in the house says, Yes, locals have been going on about it for weeks – they've been eating the croquet lawn and making a noise. The usual complaints. I drive downtown and see corellas in the dead trees (dead from salinity, however, not from the corellas, as some claim) along the river. I cross the bridge and park next to the church, as I see utility vehicles parked outside. I jump out – round the old Holy Trinity Church where I attended services as a child, struggling between scepticism and soul, and formulating my own spiritual anarchism – to see a guy with a shotgun. And a boy on a two-way. A number of guys walking around searching the trees and skies and firing.

The shotguns are no cheap pieces – they are what such shooters consider artworks. Sports shooters. I scream at one of them and call him a murderer. If nothing else, can't they even respect a house of worship? I ask if they're professional shooters, which makes it no better, but it says something about what's going on. No, is the reply. Go and see the ranger.

I do – she's fifty feet away. I yell at her and say: this is not the way, this is not the way. You are wrong. I lose my cool. She sends me to the shire. I drive down there and have an angry exchange with one of the shire officers. If nothing else, having sports shooters in town on a Saturday morning, with families and their dogs walking around, is not on. I can't make you see that the corellas matter, but surely the bloody people do? It then degenerated into what I thought of the shire and its failure to protect the environment, now and on any other occasion, and I stormed off. It was all very extreme. I probably achieved little, but coincidentally or not, I heard no more gunshots that day.

These two incidents turned out to be unrelated, but they do speak a lot of notions of the 'country'. Though I live out in the wheatbelt, and though I have a lifelong connection with it, I am not 'country'. Nor am I 'city'. These are terms used to control discourses of place, movement, and especially the production of food (and mining).

I live where I live to help conserve the 'nature' that's left. I write poems of resistance and protection. I travel to the city by train rather than by car, as often as I can, there being one of Western Australia's few country passenger rail lines close to Jam Tree Gully.

To fill in the rest of the picture of this weekend: I spent the time at Jam Tree Gully, a rocky hillside block abutting a nature reserve with acres of York gums and jam trees. Rehabilitating the land, very much denuded by horse and sheep grazing, is the aim of my coming years. Off-grid, off mains water. Over the two days, I watched and photographed mistletoe birds, thornbills, willie wagtails, magpies, a kestrel, two eagles, a mulga snake, white-chinned honeyeaters, pardalotes, an orb weaver spider, bull ants, and a family of kangaroos. I watched their movements – how local birds follow the nomadic birds as they pass through, the locals going to the edge of their own territory, then flying back; I watched how obvious edge-effects like roads and even fence lines with firebreaks work as imposition or are adapted into larger pictures of light and crossover to do with rocky ledges, gullies, and vegetation types. In watching, I understand how better to write a poetry of resistance that will declare the necessity of preserving this region. Can it operate without me shouting out my poems against the shooters, the shires? Whatever the answer is, I do know that every act of resistance adds together, and remaining non-aggressive but resolute in response is what slows the assault against the environment. The assault is remorseless.

I have not yet written the poems that go hand-in-hand with these actions, though I have seen them in my mind's eye, because they happen as I interact and respond physically and emotionally to the world around me, and also they appear between the lines in my notebook, attaching themselves to broader ideas and counterpointing received systems of thought. All my poems are a dialogue with a history of ideas and the process of literal interaction with a locality. They attempt to respect differing cultural and social registers, but to speak outside them as well.

Really, though, the activist moment that becomes a poem is often away from the incident or the moment of witnessing. It comes with the 'parafigurative' collusion of the politics of response and those ambiguous receptors of the moment when an image forms almost without logic. And that was last night.

Once again, around 9 p.m., a strange and confusing noise arose outside. I went out to investigate. As all in the house described it, it was like a mob of injured birds calling out. I thought of the corellas – maybe some had survived and were on the block calling in pain. Torch in hand, I reached up the hill; then suddenly the noise intensified and I heard a rush, and the sound of feet.

Moments later, the sound came from a different paddock. I walked over and the noise became a mixture of growls, squawks, and screams. I shone the torch in the direction of the sound and two pairs of eyes caught the light. One on top of the other.

It was foxes, mating. Foxes who'd been missed in the hunt. I turned the light off and left them to it. If ever there was a sound of pleasure and pain rolled into one.... The poetic analogy is obvious and irresistible. And that's where the poet activist has to be careful – what I can take of these is what I take from the events that precede it as well as this moment. Foxes and corellas are both considered vermin. The corellas increase in number because of clearing and monoculture. Foxes were introduced in the nineteenth century as sport. Entertainment by way of killing them is sold as environmental, and yet the pleasure is all in the hands of the shooters and those who incite them. In there is the entire politics of what I write, but also their survival and liberty.

CHAPTER 3

Why I Am a Pacifist

As a child, I was taught that violence is wrong. I was taught it in the home, and in church. My father was violent towards me; my mother never touched me in anger. (My father and mother split when I was still a small child.) I was obsessed with playing war as a child, and read military books compulsively. I liked the maths of it. I preferred enacting the side defeated, and reversing the outcome. I developed a sense of justice that meant the loser would become rightfully triumphant. All battles at school, in the home, the playground, inside myself, were fought in this vein. Brought up outside the patriarchal family unit, I searched out alternative hierarchies. Peace, as a concept, was a conflicting hierarchy. Peace was an obligation outside war, and I linked it with satiety and reward, and the outcome of the 'just' victory. I have spent my adult life wrestling with this. For me, peace is a permanent state that exists outside notions of right and wrong, oppression and liberation, the just and the unjust. The person imprisoned – 'rightfully' or 'wrongfully' – can have peace. Pacifism is the one universal that defies enslavement. Peace is its projection into the material world. As a pacifist, I am a believer in peace.

My pacifism does not come specifically out of a religion, though it may share things in common with aspects of religious pacifism. It is not the stance of a Thomas Merton who could write on 'non-violence' yet also admire and revere Joan of Arc, a military icon, as well as spiritual victor. I cannot talk in terms of the victory of the spirit over aggression, over the material. Pacifism as a unifying concept can exist in the presence of so-called 'evil'. It is *that* overwhelming. This is in the belief not that a kingdom of greater good awaits, as was the case with early seventeenth-century Dutch Mennonites, but that peace is obtainable and desirable immediately and always. In *Pacifism in Europe to 1914*, Peter Brock writes of these 'early Mennonite nonresistants':

> They were but transients in this world; their country was not a terrestrial kingdom but a heavenly realm. They held, moreover, that injustice and oppression were the inevitable lot of the godly in this life. (177)

These Mennonites did, however, confront the terrestrial issues of peace in a more direct way as time went on, according to Brock:

> But with the sect's gradual acculturation to Dutch society, and with the absorption by the brotherhood of patriotic sentiments, new aspects of the peace problem began to interest Mennonite writers. They took pains to show that hope for the achievement of justice through war was illusory. Did not the innocent in war suffer along with the guilty? Were not war's inevitable concomitants – pillage, rapine, and destruction – incompatible with the ends of justice? Could one in fact discover in history a single interest of a truly just war, one that merely began, but also ended as a just war? (177)

I take these positings from Brock as my point for departure. For me, from the absolute pacifist perspective, the suffering of the guilty is, within the terms of the language of 'justice', as unjust as the suffering of the innocent. Given this, justice can get no foothold in the material world without leading to an absence of peace. A just war is as violent as an unjust war. The idea of justice and the idea of salvation are, to my mind, obstructions to pacifism. However, as a peace activist, my intention is always to help 'liberate' the oppressed, the poor, the compromised, to allow them to enter the dialogue of difference on an 'equal' footing. A protest against oppression and suffering is synonymous with a protest against war.

Pacifism is an idea and an ideal, but it comes out of the physical, the real. It's conceptual, always vulnerable to circumstance and being tested. Maybe that's why I distinguish its material realisation as peace activism, the experiential. 'Pacifism' as an ideology or belief system has a history, and a set of precedents that make adherence to its principles more certain. It has a language of ethics. To pay lip service is easy; to live out the belief is something different. To have been violent, experienced violence, and to have concluded that it is illogical on all levels, is a journey. Pacifism becomes an end result, not a process. That is because its ideal form is a constant. Always there, even if we can't see its pragmatic worth.

On behalf of a certain idea of the pragmatic, though, the hypothetical and ridiculous question is often asked: would you hurt an animal if it were hurting your child? This is one of those non sequiturs that comes with the territory, like the did-you-know-that-Hitler-was-a-vegetarian furphy. As it happens, Hitler wasn't a vegetarian – he was fond of eating small birds – but even had he been, it would only show that dietary choices and ethical outlook do not necessarily coincide. That first question about the animal and child I usually ignore, though if I feel it's asked with genuine concern my answer goes something like: *of course*.

Any belief system that does not have its pragmatic element will not work in the material world. Vegans face this in life-threatening situations with regard to contemporary medicine. Most medicines contain animal by-products, and are tested on animals at some point or another in their development. One can easily make a conscious choice to die rather than use these products, but in life-threatening situations such choice is usually removed from the sufferer. Furthermore, doctors applying such treatments believe they are taking a life-affirming step, and are unlikely to follow, let alone entertain for a moment, the ethical regime or coordinates you have declared for your own life.

Most car tyres contain animal by-products: it is, once again, almost unavoidable if participating in the modern world to bypass the car or bus. This does not mean one must own a car, or condone such animal usage, and indeed a conscious choice can be made to minimise participation in such exploitative materialism. I do not wear clothes made from animal by-products, avoid purchasing any domestic products that contain them, and never knowingly consume them. It's a constant process of vigilance and ethics. In essence, it's this consciousness and this vigilance that are at the core of the ethical regime.

My concerns relate to issues of pacifism in two ways. The first has to do with the principle of peace: of not knowingly inflicting hurt or damage on any animal. Within the pragmatics I mention, this avoidance of harm extends where possible to plants and 'the soil' as well. The second, and of direct relevance to the conjecture that follows, relates to peace making and peace maintenance. These are activities that can generate passive-aggressive acts of 'non-violence' and of 'non-resistance', even 'resistance'. They are quite distinct from a complete abstention from any aggression whatsoever: an inflexible 'non-participation'. Maybe this is a spiritual constant that the believer in peace works towards, as opposed to the sublimated aggressions of peace activism made pragmatically necessary by the obvious and overt violence of war making. In this context, 'pacifism' is ultimately unrealisable, but a variety of peace-states-of-being are achievable.

Pacifism for me extends to all living creatures: I do not eat them, do not consciously exploit them, and do not make tangential use of them through by-products. My pacifism is synonymous with my veganism and my anarchism. The State controls through varying degrees of curbing the individual's desire for self-determination and expression. It controls the group through the same tactics.

I suppose, though rejected in many ways, pacifism became defined for me – an active alternative – out of the Christian church. Christ teaches us to turn the other cheek. I saw the church as another State, another system of control, though I could admire the figures it used to hide its own violence. The church knew that the Lamb of God, the willingness of one individual to absorb the sufferings and sins of all others, would allow it to control by diversion. In the name of...? I searched other religions too and found this inherent 'pacifism' everywhere I looked, regardless of the militaristic rhetoric accompanying battles between 'good and evil'. All religious works dichotomise peace and violence. The ideal wrestles with the pragmatic. Peace activism starts 'left of field', but must be vigilant against its own internal violences.

I admire Gandhi because of his passive resistance. But from a distance, it also made good 'tactical' sense. I ask now whether that's pacifism or another form of aggression. From a peace activism point of view, it was positive and consistent. It moved towards a pacifism. The internalising of outer conflict by converting it into personal suffering is noble and selfless, but it inflicts a form of violence on the self. This, of course, does not invalidate it, but is it self-suffering pacifism or a form of peace activism? There is no easy answer to this, as it involves questions of spiritual beliefs as well as social and political tactics, so the implications and outcomes become entangled.

I admired the Jains because they did not eat animals – decidedly. But that doesn't mean some Jains didn't make 'effective' generals. The one factor does not necessarily invalidate the other.

A search for peace? Or ways to think about peace? An act of appropriating – a violence in itself – other ways of thinking about cultural interiorities? Peace is not bound by culture? Or does peace come out of culture...? Peace seems relative. Pacifism linked to peace seems essential to me.

'Conscientious objectors'? In Australia as I was growing up (I was a child during the Vietnam conscription days, during the moratorium marches [Australia's most recent period of compulsory national service extended from 1964 to 1972 – NL]), 'conscientious objector' was a dirty expression in a society filled with uncles who were war heroes – those who'd done their bit to make 'my' freedoms. War comics, TV, war games, wars in the back yard, fights, being beaten ... aggressions: 'out of it' aggression ... psychiatric revolution ... anger. I was going to join the army as a kid. To be a general. Top of the pile. Less risk and more control. The Australian Defence Force Academy even holds some of my poetry papers – every poem an utterance against violence. Unsustainable ironies?

One of the ironies of the present 'war debate' – of any war debate – is the idea of a right or a wrong war. I was offended on hearing that Andrew Motion (Poet Laureate of the UK at the time) wrote an anti-war poem while adding that he might very well write a pro-war poem should it be proved that Iraq *is* hiding weapons of mass destruction. Aggressive liberalism and the idea of the 'just war' are at the core of the perpetuation of violence. As the coordinates vary according to social and cultural perspective, so does the notion of justice. This is a truism, but one that seems impossible for members of nation states to get around. The idea of a collective wrong, or of being collectively wronged, removes nuance and subtlety. And necessarily makes all responses insensitive and brutal to some person or group of persons at least.

Even within terms of the violence of international militarism, why should a few nations with certain military capabilities dictate the arms policies of other nations? The 'reliability' factor and the 'responsibility' factor are linked with notions of 'civilisation' and righteousness. The United States government plays this card on a consistent basis. The presidential executive becomes a secular godhead working as a mouthpiece for its phalanx of avenging angels. This is a nation that deals in euphemisms; things are never called what they actually are.

The 'enemy' will always play off this omnipotence – either through Marxist denial or disclaiming of the universal godhead, or through opposing the usurpation of God-like omnipotence by the US government, which ultimately still models itself on the Christian God, or more specifically on the God of individual salvation: a concurrent individualism of profit. As a pacifist, where can one go with this? It's often a slow journey to complete abstention from a belief in the justification of aggression in any form. The idea that history teaches us is refuted by the activities of governments now – unless it teaches patterns of violence.

How important is it to differentiate between pacifism as an ideology, and 'non-violence' as a passive-reactive tactic of confrontation? Can a pacifist utilise assertive actions, employ strategies of 'non-violence'? In asserting the rights of identity, in not being oppressed by those wishing to gain from this action, a form of resistance comes into play. Is this pacifist? For me, peace activism is a political strategy informed, to lesser or greater degrees, by a politics of personal interaction. Can the violent psychotic be a pacifist? I would answer yes, if his or her intentions were peaceful; if he or she rejected violence as a means of change or as a means of resolving conflict; if the violence came out of an undesired systemic or chemical condition. Pacifism is a constant we aspire towards. It's a belief.

However, the violent psychotic would seemingly not make an effective peace activist. I am forging meaning here. But pacifism is about intention. It's a resistance to the desire for violence – a violent person might still aspire to pacifist behaviours. If someone hits me, I do not hit back. Fact. I have done so in the past: living on the street, suffering major addiction problems, I 'defended' myself. I'd like to add 'inadequately', but that's beside the point. I 'knew' it was wrong. The moment I stopped defending myself I was assaulted less. Maybe it's a body-language thing, or maybe it's a matter of the defensive posture being an aggressive stance.

I recall as a teenager reading about a type of tank the Swiss had invented that was useful for defensive work in high terrain. That it fired most effectively from a stationary position. Defensive tactics. Even then it bemused me how weaponry could ever be defensive. An inverted sublime. Like anti-personnel mines, or anti-tank weapons. The 'defence' industry thrives on euphemism. In the end, language is the strategic ground for conflict. The non-pacifist or non-peace activist will openly use non-violent means if it is not tactically prudent to show outright aggression. Non-violence, like 'anti-tank' weapons, becomes a defensive militarism. This posturing works well for those wishing to maintain a status quo while giving the impression of pacifism. The male privileged within the family unit, the guy at work barely sexually harassing a colleague, the person privileged within a community by his or her colour.

A believer in peace might not be a pacifist or peace activist. Peace is a variable, a relative state of being. Peace, for some, might be a desirable position between nation states, between different ethnic groups, between football clubs, but not an absolute. The believer in peace might, like Kant, subscribe to national or personal defence. As W.B. Gallie observes in *Philosophers of Peace and War*:

> Although not a pacifist, Kant regarded war as the greatest evil besetting human societies, and in one passage he goes so far as to describe war as the source of all evils and of all moral corruption [see Kant, 'The Contest of Faculties', in *Political Writings* – NL]. But he did not see war as an evil which admits of any one complete and immediate cure. It was the extreme form of the general evil – the natural egoism – in human nature which had, first, to be tamed by the enforcement of laws, no matter how harsh and imperfectly rational, and which only thereafter could be directed towards the political ideal of lawful freedom, within which pure social morality – men treating each other as ends, never as means – would be at least partially realised. But while insisting on the inherent evil of war, Kant acknowledges that every citizen should be prepared to defend his country from foreign invasion. Indeed he seems to have regarded self-defence (at the national as much as at the personal level) as a natural reaction, essential in life. (21)

Gallie's interpretation of Kant's views is useful in establishing distinctive correlations between the 'abstention' of absolute ideological or 'spiritual' pacifism, and the necessary pragmatics of peace activism. As a pacifist, I do believe that war is a 'wrong', but I also agree that it's part of a 'natural egoism': that is, that there's an inevitability in the human condition that leads to disagreement and conflict. This is wrapped up – on the surface – in issues of freedom and persecution, power and disempowerment, but is chemical, structural, and spatial as well. However, my pacifist means of resolving the tendencies of 'personal egoism' differ greatly from those projected by Kant. As an anarchist, I resist the State's imposition of 'law' to control others or myself, and I emphatically believe in a self-imposed ethical 'law'. Defence for Kant becomes an act of physical aggression, as much as of resistance. It participates in the language of war. I favour peace activism as mentioned above. Furthermore, I see 'nation', as in the 'nation state', as being a machine or device of war.

I am a pacifist. This does not mean I subscribe to actions of 'non-violence' or 'non-resistance'. These imply a language of violence. I would prefer to work out of a literal language of peace. To dialogue – a catchword of contemporary internationalism, of peace activism. Also a catchword of globalism and economic duplicity!

My poetry has often been critically described as 'pessimistic' or 'dark' or 'full of violence'. This is poetry of observation and experience, which does not prevent it from also being a language of peace. It's a form of peace activism that comes out of pacifism, and in the extreme, not a pacifism itself. If I subscribed to my 'constant' view of pacifism, I would not write.

I feel poetry is not a telling; it is an evocation, a suggestive disturbance. But through such disturbance, a possible peaceful environment is perceptible. The actuality is not the potentiality. This brief essay is another step in a personal struggle towards the articulation of seemingly irreconcilable beliefs in equality and freedom, and pacifism.

Pacifism – peace in all interactions (gender, domestic, spiritual, etc.) – cannot oppress or predate in itself, though an idea of pacifism might be deployed by an aggressor or passive aggressor as a means of empowerment. But in itself it is inviolable, by definition. By forming a constant.

A peace activist might work every method of non-violence and, in the end, enlist in war if his/her community is directly affected. A pacifist will not go to war, regardless. A pacifist might react at the spur of the moment to defend a loved one – a pragmatic spontaneity – but will not consistently make the choice to do so. Defence can be an involuntary biological reaction. The peace activist who believes in the constant of pacifism will attempt to revert to 'non-participation', to total non-response in the moment of personal crisis.

There's a fascinating outline of some issues regarding 'struggles' internal to peace activism in Sasha Roseneil's *Disarming Patriarchy: Feminism and Political Action at Greenham*. In many ways the success of the Greenham Common activists' commitment to maintaining a primarily peaceful protest against the presence of US cruise missiles in Britain is a reference point for assertive, primarily 'non-violent' action or protest. However, it would be naive to think that issues of conflict on an intrapersonal level and within or between groups didn't take place. That's the nature of what Kant called 'natural egoism'. It does not necessarily invalidate the action, though. As Roseneil writes:

> Some argued that only women should take part in actions to ensure that they remained non-violent. Experience of the violence and aggression of some of the men who had been living at Greenham suggested that the women were less likely to be violent. Some also believed that the police were less prone to using violence against women protesters, and therefore the camp should be women only in the run-up to the threatened eviction. (40)

It is interesting to note the 'were less likely' here. This is not a denial of the potential violence of women, but a comment on the discourse men are raised in. I have experienced political violence from men and women, but have most often found a readiness on the part of men to exact political revenge in an overtly aggressive way. The second point, regarding the reduced likelihood of police being violent to women, is relevant in the same way, though I note from personal experience that a greater number of female police will lead to greater violence among policewomen and other women, as well as from women to men. This is a tactical issue once again, and not an issue of pacifism. It's non-violence as direct action.

I have been at many protests where things 'turned nasty'. Hyped-out on anger or the occasion, or a mixture of these and substances, the most peaceful intentions can become more of the very thing under protest than the peace-making beliefs that produced the original action. The extreme expression of the pragmatics of pacifism, via peace activism, is peace enforcement: the aggressive protest, or the peace keeping troops of the UN. For me, this is one step too close to the cause to be truly effective. However, it is important not to debase intentionality with 'errors' or diversions in the peace activism. A protest of ten thousand people with only a dozen violent participants is not invalidated by the negative activity. The group is an effective tool for change, but individual responsibility is its own cause and effect. The whole thing takes a working combination of both.

Roseneil continues:

> Others, more explicitly feminist, took the position that for only women to deal with the authorities was more than a tactic to avoid violence; it was a symbolic action drawing attention to the gender politics of militarism and its support from the state. This was often coupled with the argument that working within a women-only environment was empowering and would build women's confidence and strength. Finally, others argued that men were a drain on the camp, had not taken responsibility for their share of the domestic labour, and were pushing women back into traditional housewifely roles. (40)

There are two issues of passive-aggressive violence raised by this passage. The first revolves around the gender identity and militarism vis-à-vis the state, and the second, in terms of domestic hierarchies, again around gender. The refutation of gender tyranny, especially with regard to inequalities within the domestic environment, is most often expressed through subversion, through verbal or behavioural self-assertion, or specific acts of non-violence. The wrongs of hierarchy are obvious. The gender conflict is resolved within the pragmatics of peace activism. It is not a pacifist action, though individuals may have been pacifist. The issue of group dynamics is separate from the ideological constant of pacifism. It's a passive-aggressive action. Can the tactician remain a pacifist in this environment? Ideologically, is this a resistance to the idea of violence, or simply another form of it? Maybe the end results could be seen as 'just', certainly from a feminist perspective, but are they just on all sides? Pragmatically, it seems to have been a positive move and to have led tactically to increasing success within a context of peace activism. It's of interest that, according to Roseneil, 'The men who were asked to leave responded angrily, one exacting revenge by picking up an axe to chop down a shelter, another, a non-violence trainer, throwing a pot of water over a woman; but they all departed' (40).

As a child and a teenager I used to hunt and fish. I also suffered a lot at the hands of school bullies, and was even hospitalised as a result of bullying on one occasion. There's very likely a connection between the killing of animals and being bullied. It was not until I made a link that I could let go of both. I had an opportunity for revenge once – the police wanted me to name one of those who assaulted me (they were only interested in the one non-white guy, the five 'whities' were not to be named). I refused to give a name, and it showed me how a sense of empowerment can come when one sees no empowerment in a course of action.

There is something disingenuous and non-pacifist to my mind about empowerment, but it's a step on the way. I became an active protester against environmental damage, in favour of Indigenous land rights, and against militarism and nuclear energy. I was arrested on a number of occasions for that, and for substance problems. I called myself a pacifist but was aggressive in the extreme. I was angry about the oppressions and injustices of things as I perceived them. I gave up substance abuse and gave up aggressive protesting – which is not to say I won't still march or speak out for a 'cause' – I do, consistently. The difference is that I don't see my actions as tactical 'non-violence'. They are actions complete in themselves, and not intended to evoke violent response (though they sometimes do – including death threats). It becomes a matter of intention as much as awareness of likely outcomes. I realise I will upset people, but it is not my intention to do so.

In the animal rights movement, there are hugely varying beliefs over such issues. As a vegan pacifist I am ideologically against all forms of violence – including 'non-violence'. From the extremes of bombing and damaging hamburger chains and animal testing facilities, through to the 'outing' of vivisectors, a desire to take on the violence of those who damage animals is maintained. An act of revenge is a vehicle for personal anger and blood letting. The pacifist vegan, however, hopes change will come from pursuing one's beliefs without aggressive intent. My non-usage of animals in any form, to the best of my knowing ability, sets an 'example' of respect that is its own end. It is not a tactic of non-violence, as I am not trying to subvert others into thinking the way I do. I believe so emphatically that it is the right thing for me to be doing, that I do not feel the need to destroy anything in the process to make it a reality. I would die before consuming animal products; not as an act of civil disobedience, but because that's the truth to me. That's the key to my pacifism. However, while not trying to force change on my neighbour, I will march and write and speak out against the systemic abuse of animals: this is the pragmatics of it, another form of peace activism from my point of view. It becomes an issue of rights.

Is this a reconcilable paradox? I feel it is when linked with daily patterns of behaviour. One time in my youth, while hitching in the south-west of Western Australia, my brother and I were nearly run down and were actually hunted (bullets fired around us – they couldn't pinpoint their target) because we verbally abused log truckers for wrecking the old-growth forests. Our desire to stop violence against trees predictably evoked violence. Now, by contrast, I would sit down and have a yarn with the truckers and share views if possible – the key being that I would not seek them out to do this. That's part of the pacifist consciousness – allowing it to happen organically, rather than setting it up.

Inevitably, we arrive at the coming 'Bush–Blair–Howard' and all other vested interests in war. Where do I stand? Against war, all war. Against is the wrong word. 'War' is a misnomer. I don't recognise it. It's a non-state of being. It is against life. There is no such posture as the defensive. The national posturings reek of irony – linguistics tells me that. Even in its own terms, it's an 'unjust' war. One speaks about it as if it's inevitable, and yes, that's where the justice on its own terms ceases to exist. Whatever its goals, its intentions overwhelm them. My pragmatic self invokes peace activism!

In the UNESCO collection of essays *From A Culture of Violence to A Culture of Peace: Peace and Conflict Issues*, Felix Marti, in 'Understanding and Dialogue Between Religions to Promote the Spirit of Peace', calls for dialogue and peace in place of violence. He explores the problems of religious violence with regards to (a) god's expectations of justice and then makes a point about peace being a 'consequence of justice':

> Peace is a consequence of justice. In situations where people or groups are oppressed, it is normal that there be violence and wars. Human history shows a series of violent situations because injustices have consequently taken place. The change following the advent of modernity is that injustice is no longer considered to be a consequence of fate, so much as the result of decisions by humans and of structures that can be modified. Freedom and oppression, prosperity and poverty, are social constructs, and violence therefore arises not only as an expression of protest but as a way of breaking down unjust structures. In our age, the desire for justice is now universal. Before, there were above all local claims which aimed at a higher degree of freedom or fraternity within geographically well-defined societies. Today what is really shocking is the injustice that condemns many people to live in conditions inferior to those of others and with a very limited level of freedom. Justice must become global. There will have to be changes in the political and economic structures that ensure prosperity for a minority of nations and maintain the underdevelopment of the majority. (148)

I feel it is patronising to those in poverty, and/or suffering oppression, to deny them the agency of peace. Violence does not bring peace, and only peace can bring equality and non-suffering. Absolute peace would deny exploitation on any level. Peace is an internal as well as a collective search, and that's the only viable 'action', unless the history of 'violent situations' is not a permanent condition. However, I feel Marti is right in recognising the necessity of resolving issues of oppression and inequality. My problem is with potential means, and notions of 'justice'.

Justice and peace can both become euphemisms for violence in this context. An entire shift in reasoning needs to take place before they can take on qualities that are intended in their universal usage. The recognition that power corrupts justice is one thing, but the belief that an end of oppression will bring justice, and consequently peace, is imbuing peace with a subjective (if 'good') set of values that it cannot sustain. Pacifism is, primarily, the consciousness of being peaceful – it is a given, a fait accompli. It is not a denial of choice, but a starting point from which all choice comes.

There is no freedom, no choice, in a discourse of violence. All nation states are built out of a resistance to threats against their existence. They react through localities of worship, the judiciary, the military, the media, their buildings and cultural languages – they resist dissolution through aggression at worst, cultural 'non-violence' at best. Either way, they can never be peaceful. Institutionalisation is violence. One might more realistically say, 'war is a consequence of justice'.

CHAPTER 4

Plagues & Bioethics

> But be it which of these it will, when our travellers began to perceive that the Plague was not only in the Towns, but even in the Tents and Huts on the Forest near them, they began then not only to be afraid, but to think of decamping and removing; for had they stay'd, they wou'd ha' been in manifest Danger of their Lives.
> – Daniel Defoe, *A Journal of the Plague Year*

I

This essay has been 'evolving' over three years. Every time I return to it, another 'food crisis' is dominating the domestic and foreign policy agendas of Britain and the world in general. The difference is that now, in 1999, I am writing this from the American mid-west, from the rural heartland. I'm observing the reactions here in the USA and in Canada to the 'foot and mouth' crisis in Britain – a real conflict between self-concern and obligations to an ally, mixed in with old world/new world views on 'infection'. Gambier, Ohio – surrounded by cornfields and farming communities – engenders a feeling of rural security. Yet these apparently tranquil cornfields are a laboratory, a genetically modified epicentre for a commercial experiment that has the human body as its ultimate target. It is possible to go into shops and buy what are claimed to be non-GM foods, but the prices are high. For the ordinary family, on an average or low income, genetically modified foods are a staple – and GM creeps in at every level of the production chain. Choice, in effect, is denied. In a neighbouring county, there is a strong Amish presence, rejecting the new. We wonder if the grain they grow is GM-free, original seed...?

Recently, I've had to do a lot of travel between America, Europe, and Australia. Every time I enter a new country, I'm made aware of the anxieties of quarantine efficiency. With 'foot and mouth' still prevalent in

Britain – despite the propaganda from the Blair government – and present also in Europe, there is a stronger than usual vigilance at American and Australian airports. It's the same atmosphere as when the early outbreaks of Creutzfeldt–Jakob disease were discovered in 1985. There's an irony in this – as many vegans here know, the animal form of the disease has been present in American livestock for years, but disinformation has kept it quiet. Another version of it was recently discovered in deer. One wonders if it isn't already present in Australian livestock too, and whether, as with the early AIDS cases, symptoms matching the disease are ascribed to some other illness. The various industries involved have a lot to protect.

Quarantine isn't just about keeping diseases out, protecting a specific geography from physical contamination, but also about the preservation of 'home' values. It is about a mental and spiritual 'purity'. Back home, it's another form of the 'White Australia Policy', extending from microbes to people and ideas. [Like the 'Final Solution', the 'White Australia Policy' is irreducible to a single official document. But just as the Third Reich was undeniably racist, so there is ample evidence that Australian immigration policy under successive governments from federation in 1901 until the election of the Whitlam Labor government in 1973 was openly prejudiced against non-whites. – NL] Information might be available via the Net, media, or books, but that's quite different from allowing physical presence. Australia makes an interesting case study regarding its quarantine laws and the colonial/post-colonial relationship with the 'exporter' of disease, Britain. It is interesting to note that the Irish Prime Minister, as spokesperson for one of the ultimate victims of British and especially English colonisation, called Britain the 'leper' of Europe earlier this year. The use of this word actually takes us to a new level of offence, for even a victimised country like Ireland can just as easily perpetuate the 'us and them' mentality, the instituting of a materially pure, and spiritually sanctified, centre. No room for corruption here. The furies, whatever their religion or economic views, will have vengeance on those who pollute.

It is worth considering the cultural politics of 'plague' diseases in farm animals – the implications of 'mad cow' disease, 'foot and mouth', the potential destruction not only of Australian rural industries, but also of their self-conception, their 'way of life'. These are diseases of an old world order, of a corruption and decay that Australia feels it might still keep out. There's paranoia running through this, and guilt. Australia, as a nation, is in itself a corruption of the very quarantine laws it has developed and enforced. It is the construct of intrusion and pollution: colonisation. In the same way that many non-Indigenous Australians still refuse to accept collective and 'historical' responsibility for offences against Indigenous peoples, so, too, do they work overtime on preventing 'offence' against the sanctity of their living conditions, of their spatial and metaphysical security. The brutal incarceration of 'boat people' is a case in point. The fear is manifold: the risk of bringing 'exotic diseases' into Australia only a peripheral one. The treatment of people as polluted herd animals is strikingly obvious if you've been outside the country for a while. They do it here in the USA as well, and in Britain. It becomes an issue of power and the protection of privilege. That's what quarantine is. In the case of the animals, their well-being is not an issue. Blair ordered the destruction of thousands of animals, many not contaminated – preventative action. It's genocide, if we consider animals as beings rather than things.

The argument becomes one of a culture of meat eating against a reliance on alternative protein sources; disease is less likely to be transferred through plants. We see a rush of new vegetarians and vegans on Internet chat boards around the time of food-source crisis – 'not only the fear of disease, but seeing those animals slaughtered like that drove me to it...'. It's a knee-jerk moral reaction which, at least from my point of view, is good if it contributes in any way to developing a consciousness of the issues involved, if it leads to less exploitation of animals. Unfortunately, when the crisis 'passes', that is, when the media lose interest, or various power structures manage to 'contain' the situation, these temporary vegetarians revert to their old habits. Memory is short and selective. The problem remains.

So we need a larger cultural analysis of what it is that addicts societies to problem food sources, how the discourse of 'meat' operates in the way we see power, gender, and ethnicity. Carol Adams, in her brilliant book *The Sexual Politics of Meat,* refers to animals as the absent referent when meat is eaten. The process of animal exploitation is so entrenched that it offers a vehicle for exploitation among humans:

> In this metaphorical sense, the meaning of the absent referent derives from its application or reference to something else. As the absent referent becomes metaphor, its meaning is lifted to a 'higher' or more imaginative function than its own existence might merit or reveal. An example of this is when rape victims or battered women say, 'I felt like a piece of meat.'... The animals have become absent referents, whose fate is transmuted into a metaphor for someone else's existence or fate. (53)

The Black Death of the fourteenth century was known as the *prima pestilentia,* and the AIDS epidemic is often referred to in such alarmist terms. But in many ways the new *prima pestilentia* happens in the laboratory, behind the scenes. The modification of genes, the infiltrations of animal, human, and plant DNA, open whole new territories for occupation and exploitation – creating new dimensions for contamination in the organism. The contradiction inherent in this, in terms of the wars against existing diseases, is absurd. The Australian government, for example, will allow trials of genetically modified grains and yet fight to keep out 'foot and mouth'! When it's said there's little evidence for GM side-effects, consider the case of Starlink genetically modified corn in the US and, though approved for usage as animal feed, its contamination of human food materials and the medical consequences. The weakened organism becomes the potential host for new viruses, new mutations.

II

When I was asked originally to write this essay on bioethics, we'd just got back from driving out to look around the historic town of Saffron Walden in Essex, England, which proved as much as anything else a tour of both government and private test sites for GM crops and other kinds of genetic experimentation. Not far out of Cambridge, the Wellcome Trust has established a large facility which had featured heavily in local papers recently because of attempted expansion, opposed by ordinary residents. As with most such problems, the government publicly supported only limited expansion while in real terms overturning obstacles in order to *maximise* that expansion. This seems typical of the whole government approach to biotechnology, mooted as an area with potentially great profits, which Britain cannot 'afford' to miss out on. The government, then, will sidestep public protest in order to secure long-term financial benefits. The experimental data that might contradict apparent benefits of particular biotechnologies are either suppressed or ridiculed.

Now we don't yet have 'proof' that GM foods are dangerous, but why do their development and use proceed without proof that they are safe either? And in Britain at least there's evidence that the consumer doesn't want them, if that is the 'bottom line' it's often made out to be. We are told we are ignorant of what genetic engineering is about, that it's been going on for years, that it's no different from crossing a peach with a plum and so on. We are made to feel like Luddites (one of those how-to-end-the-conversation words – who wants to be seen that way?). I'm quite prepared for people to think my sense of something hideous when travelling past GM-crop fields is a romantic reaction, purist nostalgia for a supposedly wholesome past-that-never-was. I'm used to that from being a vegan (you will only see us represented in the media either as a social problem or as harmless, misguided hippies – at best) and from being a feminist – there will be wilful misconstruing of anything I try to think or say about these things. Yes, something at the gut level bothers me about tampering with the genes of living organisms. But what bothers me most is that it's done for profit while putting life and/or health at potential risk. It offends my sense of ethical veganism, moreover, because animals too are being fed these substances, and because they are seen as fit material upon which to experiment (e.g. transgenic pigs).

How do we go about sorting out the levels of ethical difficulty here, and which do we prioritise? I see my veganism as being at the base of my approach because it concerns how my body (which is the only thing 'mine') interacts with the world – which is not to say that is the only level of action.

My veganism is inseparable from how my body interacts with the world, but it is also a locus of logic and spirituality. I see no choice in this; it's something I have to do. When at the age of 22 I decided or discovered that to make any other life choice was cowardly or erroneous, given my social and cultural circumstances, it became blatant hypocrisy to go any other way. In a sense, the choice was made for me. I cannot cope with guilt, and transgression brings the most debilitating form of guilt. Bioethics is fused with this.

Logically, we can choose not to alter the biology of an organism, in the same way that we can choose not to eat animals or use animal products, certainly given most social and cultural conditions. You may be alienated or ostracised by your cultural group, as I have been at times, but the choice is there. Of course, some environments shape the conditions of choice in a way that excludes such options, and I respect and accept that. Veganism is not a dictatorship or a tyranny of spatiality and choice.

In the case of GM, however, it is different – education, class, wealth, and cultural priority favour those making the choices. Scientists and politicians control the flow of information, and are themselves controlled by governments, bureaucracies, and corporations. The most altruistic scientist works within a regime established for the purpose of profit or nationalism, or of shaping humanity in a particular image according to a particular agenda. We all have our prejudices when it comes to responding to acts affecting others, especially regarding decisions about what's for the common good.

To tamper with food sources in regards to living organisms is to deprive not only the organisms themselves of their rights, but also to colonise and control people on the most basic and fundamental level – in terms of what they eat and how they eat it. As for prioritising, the attack on what constitutes the individual organism is the key issue. What people eat and why they eat it will always be determined by backgrounds, cultural environments, and so on – only the eventual collapse of the meat and flesh industries will change habits on a large scale (and it *will* happen) – but most people baulk at interfering with the life signature of a creature, or a plant for that matter.

People talk about plants screaming – well, I don't eat animal products for two reasons. One, because I think it is 'wrong', it goes against my sense of justice, and secondly because I can survive and even flourish without animal products. But I simply could not live without plant products. All the same I respect the right of plants to flourish and to have their destinies determined without direct intervention into their physical makeup.

The hybridising of plants is troublesome enough – I always look for 'original seed' – and much of what we eat has come out of this process, but that is quite different from actually getting inside the genetic code of a plant and depriving it of 'choice', of its rights to identity. In the same way that meat eaters differentiate themselves on issues of cruelty – few would enjoy eating flesh that was procured through obvious torturing or suffering, despite all industries of death causing pain either subtly or overtly – so should omnivores, vegetarians, and vegans consider such issues in relation to plants.

I believe in respecting the material and non-material rationalisations of existence – my ontology consists of respect and a movement through space that recognises boundaries and shapes. Some of these boundaries are liminal and can be traversed without offence. Transgression can even be positive if such an act breaks a cycle of oppression and bondage.

Fundamental to my ethics is that doors are kept open for dialogue. I respect difference and encourage opposition. My certainties are only of myself.

CHAPTER 5

Refugees & Australia

Here I will be primarily subjective. I am not so much going to explore the obvious violations of human rights by this retrograde federal government [the conservative Liberal–National Party government in Australia, led by John Howard and first elected in 1996 – NL], its overt campaign against refugee 'incursion' described and held accountable in such works as *Dark Victory* by David Marr and Marian Wilkinson, but rather to talk about complicity. I feel most Australians, including myself, are complicit in this outrage – maybe even those among us campaigning for or advocating refugee rights. I will argue that, as in other Western democracies, the rights of the majority are used to deny or minimise the rights of minorities, and that this is a false democracy, in which rights are displaced and disguised under euphemistic terminologies. I should point out that looking at the language used by some to describe people of 'different' ethnicity will mean having to refer to repugnant and sometimes even hateful words, but certainly I intend no offence by doing so.

The irony of the increasing move to a monolithic federal power structure is that the so-called majority, those for example who voted the Howard government into power (or would have voted an equally problematic government under ALP Opposition Leader Kim Beasley into power), minimise their own rights by denying human rights to others – in this case, refugees. Precedents are set that are eventually internalised – it is easier to deny when an official form of denial is already in place. Resistance to imagined incursion becomes fuel for the repression of those already within the island fortress. Examples of similar ironies in the other places I live are the USA Patriot Act and the zero tolerance policy of Blair's United Kingdom (normal bus shelter seats removed in order to prevent 'beggars' sleeping on them, banning the sale of spray paints to under-18s to prevent graffiti, etc.) – measures designed to 'protect' or 'improve' life within a group, which may be turned upon the very members of that group. To give another example: with my partner Tracy Ryan I wrote a play a few years ago, called *Smith Street*, which satirises the extraordinary 'temporary' powers of Premier Richard Court's Liberal–National Party government in Western Australia – powers allowing for, among other things, women suspected of prostitution to be body-searched, such that by definition all women were 'suspect' and under scrutiny, and therefore also had their rights eroded. Each of these examples concerns the State declaring its intent to protect its citizens while potentially infringing on those very citizens' rights.

For me, bigotry against 'outsiders' is inward-looking in terms of nationhood; it's how a nation constructs itself by exclusion – the much-discussed 'Fortress Australia' mentality, for example, with all its subtexts. ['Fortress Australia' was a term used by critics of the Howard government's escalation of coastal border patrols and other protectionist measures, including detention camps, designed to make Australia a stronghold against 'invasion' by others. The term remains current under Kevin Rudd's federal Labor government, however, elected in 2007. – NL] And when a nation does allow immigration, it is for its material well-being, or to create a social cohesiveness (post-war British immigration), or to create social stability by relying on family structures to maintain conservatism regardless of origins. What nations can't countenance is a loss of control over these factors. Now, the refugee is the antithesis of such controls. Movement of refugees across borders in times of war is feared because of financial and social pressure on national infrastructures. In the case of Australia, with its intact coastline, this fear takes on a variety of meanings which I will touch on shortly.

My prime concern is to look not at the obvious problems with government policy and the individuals who impose it, but at the culture of exclusion that is drummed into us from early childhood. This is directed not only at so-called 'native born' citizens, but also at 'migrant' citizens who have gone through official channels, especially where they have initially suffered prejudice for being seen as intruders in their new land. There is sometimes a kind of reformist zeal, where the pain of this experience is displaced onto those perceived to have circumvented official processes, when in fact those forced by whatever means to uproot themselves from their homeland will have inevitably suffered great hardship.

For me, prejudice against refugees, whatever the reason for their movement, has always been an issue of racism. As a kid, I saw the 'Pommy' or 'Eyetie' migrant being persecuted at school, in all those familiar ways. The primary school I attended from the end of the '60s to the mid '70s, Brentwood Primary School in suburban Perth, Western Australia, was not far from one of the hostels where new migrants were housed. The litany of bigotries is a familiar one, from different accents to funny tastes in food. I felt akin to them because I carried the tags 'Poofter' and 'Dictionary' from an early age.

The conflation of sexual identity and an inclination to reading sensitised me to the vagaries of words, and their power to be manipulated. A positive word could be made 'bad', a pejorative could be invested with an entirely different meaning. The 'poofter' tag came from a single moment in Grade Three when I told the class that I was really a robot, and that only my friend Graham 'could turn me on'. Like a dam wall breaking, there was an initial giggle, then a flood of laughter. So I became the class 'poofter'. While I can see the element of humour in this, it didn't stay innocent at the time – it became something darker and beyond the initial joke. I should add that it is not a tag I am ashamed or afraid of, but it was not given as an affectionate nickname.

What being a poofter entailed, nobody was completely sure, but it was definitely not manly and certainly meant that I didn't like girls. And both girls and boys taunted me, though some secretly came to ask 'the Dictionary' for help with their homework. I was at primary school during the Vietnam war. There was one Asian kid in the school that I can recall, though this would change with increased Asian migration under the Hawke government in the 1980s. The kid was Japanese and his father worked in mining. We befriended each other, and I got to ride his dual-shift ten-gear dragster. I was told by the other kids that he was a 'Jap', that he was a 'slanty-eyed yellow bastard who couldn't be trusted', and that 'my grandpa has a samurai sword he got from one of them little bastards after he shot him'.

It was the White Australia Policy in action. Interestingly, this level of bigotry was extended to an American student whose father was also in mining. Both he and the Japanese student were seen as spoilt, and both had tastes that were threatening.

Brentwood was a State Housing Commission suburb, and renowned for at least one mass murder – an entire family shot by the father in, I think, the 1950s. It housed working-class whites in the main, and had its own police station. Next to Brentwood was Mount Pleasant – a middle-class professional and semi-professional neighbourhood of primarily brick houses, as opposed to State Housing fibro-asbestos. There was also Booragoon, a new suburb that would gradually become a comparatively wealthy middle-class suburb, with a strong south-east Asian presence. The conflicting and changing demographics drove the prejudices of the children in the school, who became mouthpieces for the frustrations of their parents. Our house was three houses into Mount Pleasant. I crossed the suburban border I did not recognise every day I walked to school. Sometimes kids from Brentwood would come to our house to hang out while their father sobered up. I knew that a few doors away on the Mount Pleasant side of this invisible divide, the same kind of thing was happening. Borders seemed an absurdity to me. Like many others, I was a poofter on either side of the border…!

I mentioned the Vietnam war just now. My Japanese friend was generally described as being one of 'them'. After all, 'they' – Asian people – Vietnamese, Japanese … were all the same. He bore the weight of their anger. A few of the kids' older brothers – even the odd father maybe – had been called up, and someone had to be made to suffer for that. Preparation for the bigoted 'resistance' to those who would eventually become known as 'the boat people' was written into their vocabulary from an early age. And their fellow student, the Japanese boy, would be recalled as a symbol of their loathing. He would become a cipher. Along with the non-Anglo-Celts, the greatest persecution was directed towards Indigenous peoples in general.

I would see more of this in high school – in both the city and the country. It was usually manifested in gang violence and overt harassment. When I think of the agenda behind the work of a revisionist historian like Keith Windschuttle, I get extremely angry – his efforts to 'correct' the record regarding 'settler' brutality (which he would have as 'alleged' at best). [Windschuttle is a leading protagonist in the Australian 'history wars', arguing notoriously in *The Fabrication of Aboriginal History* that the historical record does not support claims of widespread systematic brutality against Aboriginal people under British colonial rule. For a collective rebuttal of Windschuttle's position see the edited volume by Robert Manne, *Whitewash*. – NL] I have witnessed organised violence against Indigenous students and locals on a regular basis. I could give specific examples of fights at suburban discos, attacks with lumps of wood and metal on Front Beach in Geraldton, in northern Western Australia, and even the assault of a young Nyungar boy in police custody in Fremantle lock-up during the mid 1980s. I am a witness, and sadly I am sure I'm not the only one. None of this is idle conjecture.

Another conflation: the migrant outsider and the indigene. One lays claim to the land by arrival, and the other by mere presence as reminder of intrusion and dispossession. Both, as a consequence, are made pariahs. In the schools I attended as a child and in my youth, there seemed a need among students to declare their rights of presence, to claim an inheritance through the denial of others' rights. This was conceptual and concrete. It extended to religion (in the Brentwood school the bulk of students were Anglicans – Catholics were separated off for religious instruction and there was no room for other religious beliefs) and constructed itself in the economic. Most boys played 'wars' at school, and the 'gook' was always the enemy. The girls also played their obsessive persecution games, of being 'villagers' terrorised by a monster-outsider – the 'fat girl' – or kiss chasey in which the 'uncool' boys were mocked and called 'dirty' and said to be 'full of germs'. As someone put it to me recently, all children have their malicious side, but these games are catered to in so many ways by the education system and beyond. The language of white Australia was standard. Bizarrely, if Euro-scenarios of the Second World War were played out, there was always competition to be the Germans. I found this also to be the case in my late teens when I played simulation games: people hungered to play the Germans. Chips of Waffen SS were moved around the map of Europe without thought for the moral implications. Rarely did anyone push to play the Japanese. This has something to do with proximity, since the bulk of Australians fought in the Pacific theatre, but it's also about origins.

Prejudices against Europeans were fewer than against people from Asia, the Middle East or Africa and so on. Race was taught in terms of the Christian pyramid of being, the food chain, evolution, and any other hierarchical system it could be secreted into. We were schooled with subtexts of sexual, spiritual, and ethnic bigotries. During the first couple of years at high school, my closest friends were Chinese Australians. These migrant kids were high academic achievers, and I heard the stereotype repeated among teachers, as well as among kids, that 'they do well because their parents push them so hard'. I did well also, as did other Anglo-Celtic kids, and our parent/s were never mentioned!

At around 20 I was strongly involved in campaigning against Jack Van Tongeren's neo-Nazi Australian Nationalist Movement, based in Western Australia. Asians in general were targets of this group, which committed numerous acts of violence in the name of their cause, including the bombing of suburban Chinese restaurants. [Van Tongeren eventually served over a decade in prison for arson. – NL] My friends and I would determinedly remove their racist posters from lamp posts and buildings. The 'boat people' were targeted by them in particular. Once at an anti-nuclear rally I stood beside a woman who had her foot broken when one of the racist ANM 'soldiers' at the rally stamped on it deliberately. They were always there at anti-nuclear rallies in support of US troops, whom they saw as a bastion against potential Asian 'invasions'.

It's a tangled picture I am weaving. I want to jump a decade to the mid '90s. My family and I leave Australia to live in Cambridge, England. Our daughter suffers a few taunts in pre-school about her funny accent, but it's a 'diverse' school ethnically, so this passes quickly. Within four years she has 'assimilated' – whether by habit or design is a moot point. She has an English accent, and thinks in many ways like an English school kid – regardless of ethnicity. She has absorbed general cultural traits, and been programmed by the Church of England within her schooling environment. When she returns to Australia for six months' schooling, she suffers taunts of a more determined nature for being a 'Pom'. When she then goes to live in America she is teased for being an Australian.

This harassment, as she gets older, can be quite vicious. I won't go into it here, but the milder moments would include simple verbal denial stuff like 'they don't have radio in Australia', and so on. In her American school she learns that America had to attack Iraq because of the Twin Towers, and that Saddam and Bin Laden are close. She learns that Americans protect the free world. Australians aren't part of the picture. On American television Howard is seen rarely, and is always presented as a supporter, without agency. Australia is part of the 'coalition of the willing', but only just.

For her, nation becomes a hindrance and a source of confusion. Flags represent taunting and hurt. She has learnt that people 'defend' these flags without thought of the hurt that such defence might inflict on those who don't share such certainty in the sign. She has learnt that the sign is adapted to the occasion, and the values they constantly espouse when talking of the flags aren't as solid as they'd have you think. She knows how to draw the flags of Australia, Britain, and America perfectly. She's heard the claims made for each, and she knows about national sovereignty. She also knows nation means war as much as protection, and it means hatred as much as pride.

She is a child for whom 'nation' is in crisis. And it is the same for me. It is my strong feeling that the bigotry extended against refugees making their home in Australia comes out of both a deep conviction in a certain idea of nation, and a deep desire to quarantine a version of nation that is rapidly becoming outmoded. In the same way that multiculturalism has been reduced to an historic moment of government policy, so have claims of 'threat' to the integrity of nation become a trope.

I arrived back in Australia after a longish absence during the *Tampa* crisis. [The MV *Tampa*, a Norwegian cargo ship, rescued some 400 Afghan refugees whose boat was sinking in international waters off the northern Australian coast in August 2001. Threatening to confiscate the ship and imprison the crew if the *Tampa* entered Australian territorial waters en route to nearby Christmas Island, where the refugees wanted to be taken, the Howard government – re-elected only a few months later on the back of its 'tough' anti-immigration laws – caused a furore that spread well beyond the borders of Fortress Australia. In 2002 the *Tampa*'s crew received a United Nations award for following principles of safety and good will on the high seas. – NL] I was appalled not only by the actions of the government and military, but also by the Labor opposition's attempt to gain electoral ground by evading the issue. It is a moment in time equal to the passing of the White Australia Policy through parliament in the early days of Federation. The propaganda did its job, mixed with the natural vanity of the human character. 'Legitimate' migrants were shown objecting to refugees because they themselves had had to go through legit channels, so why not the refugees? 'No one gets anything without working for it'. The spectre of terrorist infiltration was bandied about. All I could think of was those kids I went to school with, who are a voting generation now, with kids and houses and a view of Australian integrity. A quarantine phobia. The spraying of the planes before they come into land.

For the Australian government, and many if not most Australians, the word 'border' is lost in the physical isolation of the island continent. Its unique wildlife and fauna, destroyed and disturbed by land clearing, salinity, mining, logging, shooting, etc., at such a devastating rate, is cited as a reason for the intensity of quarantine laws, of the dedication to keeping the unwanted out.

Now, I am not suggesting diseases and noxious weeds are welcome, but I do object to the mechanics of that policy being extended to people. The language of quarantine becomes a language of confinement and repression. The refugees of the post-Vietnam era are seen in terms of quarantine risk, in terms of keeping the diseases out. Thinking over refugee crises pre-Vietnam, though confronted with social prejudices on entering Australia, those displaced by the Second World War or by Soviet expansionism dealt with a different language of alienation. In the same way that American and British soldiers are 'murdered' in Iraq while Iraqis are 'killed', euphemism, distortion, and a language gleaned from the US military by CNN, Fox Media, and all their Western cohorts are deployed against refugees.

In a time when transparency is declared vital, where the airport becomes the focus of scrutinies of various kinds, that veritable symbol of Middle Eastern otherness – the veil – seemingly becomes the antithesis of nationhood understood in Western terms, the opposite of transparency. But it's a selective kind of transparency: those whom the government wants veiled in a different way, remain so. Western dignities are generally preserved; non-Western social and religious practices are transformed into signs of a different kind.

We hear the multi-ethnic nature of Australian society touted on a regular basis. However, in its national manifestation, in terms of projections of power, society is mono-cultural, as it is effectively if not officially monolingual. The fact of communities retaining their own birth-languages – spoken and written – does not signify diversity within the idea of nation. Nations allow this in order to contain it: permission being 'granted' in exchange for a loyalty to the language of power. On Social Security and other government forms there may be a dozen languages, including Vietnamese and Khmer, languages of refugees of past times, but this is only a concession. That old cliché: a privilege, not a right. They don't speak any language other than English when the national budget is being done.

The myths of nation are firmly lashed to monolingualism. So one can have a diverse range of ethnic minorities, yet their power is hedged and contained not only by government policy, but also by circumstances of language. As a poet, I consider it imperative that I undo the strictures of this 'standard' English, even while working within it. The figurative becomes an agency, a resistance to the rules and regulations that have become a constitution of denial. The real constitution of Australia is written into immigration laws, into customs regulations, into all those legalities that govern our movements in and out of Australia. Australia is defined as a nation not through a sense of people, but as a set of containments, coastline co-ordinates. It suppresses its Indigenous people to avoid dissolution of conceptual borders within (or the establishment of a concrete border such as a separate Indigenous state within), as it suppresses those who might try to cross into its space without abiding by the rules that define its being. Human rights, human dignity, never have had and never will have anything to do with this equation unless the model of nation itself is questioned, challenged, and changed.

One of the most disturbing statements I hear regularly from Australians is, 'but we feel powerless'. Well, Australians voted this current federal government into power, and after the glaring horror of the *Tampa* incident at that! Yet in a sense, this is almost incidental: for all its dishonesty and indifference to human rights, for all its barbarity and duplicity, this government isn't the root cause. It goes deeper than that. There is something awry in the way Australians perceive themselves as separate, different, lucky or unlucky, part of God's own country or forgotten souls at the end of the earth. Australians, whatever their derivation, are people. They are of the family humanoid, and have a responsibility to that family on a singular and collective level. This is a choice all Australians can make. Votes aren't only made at election time; they're also cast in the schools, the homes, the playing fields.

Living in Britain over the years during and following the wars in the Balkans, I have noticed that a very similar language of denial and alienation has been deployed by the British government against refugees from various ethnic groups of that war-torn region. Detained, refused, referred to as mafia and criminals, with residents angry at them being located nearby. All of these poured into a poisoned melting pot in which people smugglers (and their victims – dead in containers, drowned at sea) are conflated with the people they exploit. There is a conscious confusing of the codes, so refugees who have made use of the smugglers' services are suddenly invalidated.

I've always found this particularly strange in a place like Britain, where England is fighting to keep the union together, when its component parts struggle to find independence, declare their own rights of denial. It's a tangled web. Borders always mean people are losers, whatever form they take. Borders are controls over the movements of wealth, and wealth is what the wealthy want to keep in. Intrusions dilute that wealth, or water down the control of it. Wealth will leak back to those regions from which the refugees have been forced, or have chosen, to evacuate – money sent home to those still suffering or with much less. This is, of course, the same with both 'legal' and 'illegal' migrants. The wealth of nation is diluted. The British writer Jeremy Harding has explored such issues in his book *The Uninvited: Refugees at the Rich Man's Gate.*

Governments attempt to change the meaning of 'refugees'. Instead of people searching for refuge, shelter, they are seen as being in full control of their prospects – of seeing more than they claim to. An incident of oppression that drives a vast number of people to move elsewhere – a war, a drought, a natural disaster – is thought to become resolvable when the moment passes. An oppressive regime may be seen as a cause to drive one's own nation to war, but is not also recognised as reason enough to create refugees with claims to asylum. The question of legitimacy is raised. Some are legitimate, others are not. The claim of the legitimate refugee becomes contingent on the non-legitimate. One's claims are dragged down by the other. This dissembling, linked with the notion that all must have more agency, more ability to make choices, than they are admitting, provides the nation with a language of denial and resistance. Refugees are seen in terms of military movements, whether in waves, or in the targeting of specific places (wealthier, more apparently politically and socially stable places) for refuge.

•

On World Refugee Day [20 June 2003] the weekend before last, I joined a meeting and march though the centre of Perth. An Indigenous elder gave the welcoming speech to all participants. It was an incredibly generous speech. He pointed out that he and his people have had their land stolen, but could still welcome others. He placed himself in the position of the refugees, and linked the displacement and losses of his own people with those of people interned in the detention centres. Regarding possible subtexts between the circumstances of Aborigines vis-à-vis government policy, and attitudes of exclusion, Veronica Brady makes some relevant connections in the essay 'Mabo: A Question of Space', from her book *Caught in the Draught*:

> By definition, the idea of nationality is bound up with the notions of circumscription and exclusion, with fixed outlines which define us against others. In Australia, we have also the help of geography. As inhabitants of an island continent, our physical boundaries are clearly outlined. Our developing sense of ourselves has often relied on exclusion – on the White Australia Policy, for example, and, more recently, the regulation of immigration. Despite the lip-service paid to multiculturalism, we like to think of ourselves in terms of monolithic unity. According to Mary Douglas's classification, our society is based more on the grid than the group, pre-occupied with unity and defending frontiers. However, the Mabo decision blurs the distinctions we have drawn between ourselves and the Aborigines, writing them back into a history from which we had written them out, and suggesting that our unity might not be as monolithic as we think, that the other still survives within it and, some think, may subvert our purposes of prosperity and peaceful enjoyment of that prosperity. (14)

Just before he began playing the didgeridoo, the Nyungar elder at the rally pointed out to all that though it was not an instrument of his Nyungar people, but from 'up North', still, cultures can learn from other cultures. The implication was strong that we also have much to learn from those who would come to our cultural spaces. And learning is a sharing thing. I think discussions of the 'other' are obviously relevant – there is certainly and always a fear of the unknown, of the unsubscribed – but I think Australia's, and consequently Australians', mistreatment of refugees is not so easily covered by the expression 'othering'.

What's at stake here is a refusal to challenge the official representations of nation. Australians don't only demean and treat as 'other' refugees through their apparent denial; they express a fear and a renunciation of empowerment through the inability to challenge the idea of nation. A change of government would hopefully bring some surface change – see the release of children and hopefully all others from detention – but in the end this can only be superficial. [Sadly, refugee detention centres are still in place under Kevin Rudd's federal Labor government. – NL] Australians need to consider themselves part of government. Compulsory voting doesn't necessarily mean participation – one doesn't have to be informed about issues or aware of the policies when casting a vote.

The idea that Australia is a democracy (or that the US or Britain are, for that matter) has always struck me as laughable. Australians vote to allow others to make decisions for them. It's democracy by proxy, or democracy once removed. People often say one can't be held accountable for the crimes of one's parents, or the crimes of one's community. But it's not that simple. For example, I did not directly steal the lands of Aboriginal people, but at the same time I enjoy the fruits of that theft as a non-Indigenous Australian citizen.

I am both beyond accountability, and entirely accountable. There are different degrees, but a responsibility remains to change what has resulted. My complicity comes in my participation in the results of those actions. The same applies to the abuse of refugees. I did not vote for the Howard government, but I still have a responsibility as part of the society that did.

Now, I live in three countries and spend much of my time looking from the outside in, across those conceptual coastlines, into Fortress Australia. I don't see myself as necessarily connected by the fact I carry an Australian passport, or that my extended family is in Australia, or by those childhood memories. It's because I've had a role in the life of the place, no matter how small, and that brings responsibility. I feel some responsibility, probably less, towards Britain and the United States, though in many ways both places treat me as a foreigner and, with the increased international paranoia, more and more as an alien. Still, those are constructs of nation denying me, and I reject them.

As a pacifist, I reject all forms of violence – physical as well as mental. During the Iraq war, I was made to feel very foreign for the first time. Left-wing friends in America became suddenly patriotic, bunkered down. I heard from Australia that it was the same here. That it was considered almost treason to criticise the behaviour of the nation. But my allegiance is to humanity, to life. An accumulation of legal data is not an icon I wish to worship.

Against the compassion shown by the Nyungar elder at that rally in Perth, another response to difference was revealed as we advanced down the Murray Street mall area: a couple of young white Australians (as they called themselves) yelled 'exterminate, exterminate them all!', and placed a finger to my head and pulled an imaginary trigger. They continued this baiting for some time. The police didn't stir – I can't help wondering what would have happened if the marchers had done something similar to onlookers – which of course they wouldn't. It was a peaceful march, despite provocation.

Over the last eight years, as I have suggested, I have watched Australia from without. I've made visits most years once or twice a year, sometimes for a few months at a time. I have noticed it become more conservative in government policy, and less effective in its public resistance to this. I've heard much talk about its internationalism, the respect it apparently garnered around the word with its Olympic jamboree, its being part of the modern world. Statistics for Internet connections were often quoted. This is not the way I've seen it.

From afar, Australia seems to be more jingoistic and isolationist than it was in the '80s and early '90s. There's a sense of exclusiveness, and to use that word again, denial. From America this takes on a strange face. The denial seems to have been generated out of a new Menzies-era-like satisfaction in true-blue national identity, best expressed through the bringing home of spoils by sporting teams. The prime minister presents himself as an Ocker sports fan, and is there to celebrate the victories. I say this looks strange from America because to the Americans the Australians are merely good allies that need America.

Security is something Australians should look after themselves, but they are expected to look after it through recognition of the primacy of the US. Which is exactly what Australia does. In the patriarchy of international relations, Australia is perceived as a very little brother of the United States. I've had this confirmed by the true political authorities of both the international and the regional – that is, cab drivers – many times over! Jokes aside, there is a sense of Australia as outpost, the place of Crocodile Dundee and the Crocodile Hunter. It's a place of raw materials that need shaping.

In Britain, there's a similar sense, if for different reasons. Ask the Churchill College porters – I am always fending off good-humoured jokes about being a colonial. Australia's apparent adoration of the monarchy is considered slightly humorous by many a British monarchist. When troops were sent to Timor, one eminent Brit enjoyed telling me that Australia (the entity, if not the people) didn't have a big enough boat to ship them across. Now these slights against assumed national pride are predictable and in many ways insignificant, but they do point to a difference in perception of identity between Australia and its two main allies.

Britain and America would be outraged if accused of directly meddling in internal Australian affairs, though there's much documentary evidence of this in a variety of ways over the decades. But I would argue that Australia's refugee policy is an overt example of external influence on what it is that constitutes nationhood in Australia: a wealthy Western resistance, a 'coalition of exclusiveness' if you like, to those who might take what is not considered theirs by birthright or official migration.

If Australia were to make an open declaration that all Muslims displaced by tyranny, oppression of any kind, etc. were welcome in Australia, to make it their home, it would be perceived as a threat against the West, and consequently against the integrity of the US and UK. Both these countries have an interest in keeping Australia Christian, if not white. That's not to say Christians are not treated appallingly as well, but these prejudices work in degrees. There are different types of Christians. Nationhood works by assimilating difference and denying it agency; it also works by declaring difference so the assimilation process can take place. Degrees are noted so they can be melted down. The Chinese community is allowed its newspaper/s, of course, but only so it can be measured and contained.

Some years ago I sat on a panel for Anti-Slavery International in London. The aim of the event was to bring international attention to the injustices meted out to Indigenous Australians by official government policy, and to raise consciousness of Australian national prejudice. Indigenous trade union representatives spoke, as well as others from Australia in Britain specially for the occasion. The media were there in force and leapt upon a comment made by one of us for an international boycott of the Sydney Olympics. Germaine Greer was very vocal on this point, and impassioned in her plea to the world's press to highlight these injustices. I spoke of this and the need for land rights, and as a poet read the following poem, which I quote in full because, as the Aboriginal elder pointed out at the march I referred to above, there are some things in common regarding the removal of rights and basic human dignities between Indigenous Australians and incarcerated and vilified refugees, even if the issues in other ways are very different:

To The Non-Indigenous Peoples of Australia

It's the great excuse – it wasn't us, you can't blame
us for what happened two hundred, a hundred
and fifty, or even a hundred years ago.
We didn't hunt them down and remove their children.
We didn't come in and take the place.
But in truth, that's what we're doing – all over again.
Everything we do is based on suppressing their interests.
Who'll take the blame for what's happening now?
Who'll accept that lock-ups and jails are still places of death?
Here's our chance to be different, to have a conscience,
to know the difference between wrong and right.
For it's that simple. The rest of the world
can see this – why can't we?
Let me tell a story, a story close to the bone –

about a white family that was forced to sell up
after working the land for a hundred years –
leaving it nearly tore them apart.
They'd cleared and shaped the place, it was a portrait
of themselves, they'd poured their hearts and souls into it.
On a summer evening they'd look out over
the paddocks, over the burnt stubble, over
the stands of mallee, through a flock of sulphur-crested
cockatoos, into the rich red sunset.
They left to slaps on the back and sympathy
and the words, 'It's a hard place – beautiful
but unforgiving.' Their sorrow was understood.
They were not hated for their loss.
But what if this land was them?
What if this land had invested its spirit in them?
What if the land and these people couldn't be separated,
were one and the same. That when plant grew
or animal died it grew and died in them.
That by tearing them apart we left a dead place,
a place without spirit, destroyed the reason for its being.
Until we face up to what we've done and are doing,
until we make moves to put things right,
we'll be less than a people. History for us
begins with facing up to what we are.
Two hundred years back we thought we had it
to ourselves. Now, the world is watching.

The point of the exercise was to let the world know that newspaper spreads about the beauties of the Australian centre were just one take on things, distracting from other issues. In many ways, Australia is an apartheid nation. I recall it being said by a minister of Aboriginal affairs, early in the Howard government's reign, that there was a positive side to the Stolen Generations [referring to the thousands of Aboriginal people who were taken from their families as children and brought up in orphanages run by the state and churches, where many of them were abused – NL] – the paternalism of apartheid is there as well. This apartheid is in social attitudes – such as those that led to the burning down of Aboriginal homes one night in a country town when as a youth I was working on the wheat bins. My protests had me run out of town shortly after. I was beaten in the pub by a South African on a working holiday who openly bragged in front of farmers and policemen of shooting 'thirty or more Kaffirs at a waterhole, firing AK47s into the crowd of primarily women and children'.

That was in the early '80s. A refugee of that shooting, should he or she manage to find some way of getting to Australia, would very likely be detained and persecuted. The offender drove a truck and was considered a fine if somewhat wild bloke. A sort of backbone of the bush.

As a writer I do not always paint a rosy picture of the Australia I am part of. I believe it is soaked in injustices. I write of the land and with a fascination for those who work it, but I also write of its bigotries. Here's an extract from a journal entry from a couple of years ago:

> *June 11th, 2001*
>
> 1. Roof Lost in a High Wind
> 2. The Burning of the Hay Stack
> 3. 'Laved in flame as in sacrament...' (Thomas Merton – 'Elegy for the Monastery Barn')
> 4. Truck Overturned in Fog
>
> ———
>
> In this third vol of the Pastoral Trilogy a strong consideration of 'alternative' and marginalised spaces within rural communities. Islamic Katanning. Greek Orthodox. The Italian farming communities – the racist stories I recall from childhood e.g. the red Dodge stabbing incident. So, Indigenous space usurped by the Anglo-Celtic occupation. Then the repelling by those 'settlers' of later migrants – a double-pronged occupation and rejection. *The Trial* will challenge the hegemony of the empire builders even more directly than the earlier volumes.
>
> ———
>
> On Racism and Religious Bigotry in the WA Wheatbelt.
>
> e.g. On the Brethren in Dalwallinu, Cunderdin ... by the farmers' (Anglo-Celts!) kids:
>
> 'Own it all'.
> 'Wives scrawny pasty-faced stick insects with scarves. Look smug!'
> 'Odd'.
> 'Steal trucks from wheat bins'.
> 'People speak in hushed voices'.
> 'Everybody else leaves town'.
> 'Standover merchants'.
> 'Steal trucks from the silos and threaten to put people out of business'.
> 'One way ticket'.
> 'No music, newspapers, or radios!'
> 'Not short of money....'
> 'Big expensive cars'.
>
> The wheatbelt is the bastion of Anglo-Celts resisting 'the foreigners'. They are also strongly anti-boat people, anti-Asian etc. Racism is endemic.

This journal is kept every time I return to Australia. As a writer, one thing that has deeply disturbed me is the lack of action by Australian writers on the part of refugees. Some have been outstanding in their resistance, such as Eva Sallis and Tom Shapcott, but others, especially those of non-recent migrant backgrounds, do not make it part of their writerly voice, even if they are angered in their private lives. I do not know how one can write outside these injustices, They pervade everything, even when living far away.

I am not blind to the ironies of talking of these issues in Britain and the USA, nations themselves guilty of numerous violations of human rights, even if they pretend and claim otherwise. I feel they should equally be condemned abroad, and come in for the same scrutiny, in an international context, as Australian human rights violations. The process works two ways: it embarrasses nations guilty of malpractice in their tourist and 'brotherly/sisterly' markets, and it creates self-awareness. These are universal problems, compounded, isolated, and protected by an idea of nation. The fear of economic downturn as a result of this embarrassment is one of the surest ways to bring at least cosmetic change, even if in the short run the scrutinised and exposed oppressors kick out.

During the *Tampa* crisis I sent an email to an Australian literary discussion group voicing my outrage at the way the government was behaving. I said that all Australians were morally culpable. For this I was attacked aggressively on the list by some members, and actually received anonymous death threats. Now, this is part of the dialogue that surrounds the national literature.

Here's some of that exchange, which I reproduce because it's on the public record. I begin with my original email:

> i am looking for support to condemn and pressure the australian government re the refugee crisis. howard must be stopped, and civil and human rights come through – this is a catastrophe, and another shameful moment in australia's cowardly and shameful history of 'human rights' abuse. there is urgency in this matter, given the plight of those on the norwegian container ship off christmas island. will people please email me if they wish to add their support to this protest – i intend to send letters to a number of international newspapers and human rights organisations. thanks.
>
> best,
> jk

Re: REFUGEES
Date: Fri, 31 Aug 2001 14:54:27 +0800

I feel very sympathetic towards the refugees, and as the people on this list who know me would tell you, I have a very cynical view of life.

It is my understanding that Australia has not violated any law, it's only piracy if the foreign ship is boarded in international waters.

Sharon [an earlier poster, in broad sympathy with my original post] your reluctance to take up residence elsewhere says it all, doesn't it. Australia IS the best country in the world, so why are you rubbishing the place with your rhetoric? Attack Howard, by all means if you must, but don't denigrate every man woman and child in the country with your generalisations about Australians.

X

Let's go back to the current situation. SAS troops boarded the Tampa and relieved the Norwegian captain of his command of his vessel – an act known as 'piracy' elsewhere but obviously not in big, generous Australia. So they provided food and porta-loos to the refugees. The sight of assault rifles slung over their backs as they go round the ship must be very frightening for those on board, particularly the children using those porta-loos.

As for expecting the Indonesian Government to assist, perhaps X might like to instruct those generous SAS troops to invade the fourth-most populous nation in the world and persuade the already fractured Government to take back the refugees. The resulting conflict would make the trip to Timor look like an outing to a theme park. All those Australians who are afraid of an invasion from the north would be hiding under their beds.

As for emigrating, X, this ashamed Australian tried that about five years ago and found I liked it too much back here to stay away for too long. Having seen what a good portion of the rest of the world was like, I knew that home was the best place to be. I brought a 'legal immigrant' back with me to share this great country and he agrees with me that this is the place for our children to grow up.

Those who should emigrate are those with intolerant attitudes preferably to someplace like Afghanistan where freedom is non-existent.

Y

best country in the world? what does this mean? best for whom? not for humanity (or animal life) in general. surely...? best best best... besting...? cynical or not, it's this kind of stuff that reinforces and perpetuates the exclusiveness and aggression of nation and nationalism. there are good writers on this land mass, 'good' for all sorts of reasons, but their being 'australian' isn't reason enough to celebrate their achievements. any more than it would be if they were 'american', 'british', 'french', 'japanese', 'indian' etc. as for the SAS – just a bunch of trained-up killers, however you look at them. military is military. the 'elite' is supposed to invoke pride? in what?

the white australia policy rolls on and on and on, adapting to the times.

thanks for the support I've received re a letter of condemnation of the howard government and those supporting its present stance.

best,
jk

Subject: australia's shame – poem
Date: Sun, 02 Sep 2001 14:05:49 +0800
From: john kinsella kinsellaj@kenyon.edu
CC: Austlit austlit@lists.vicnet.net.au
References: 1

Teleclogue

Anchorperson

The invasion of boat people
needs to be contained, the flood,
tide that needs stemming,
invasion threatening our
way of life, pictures just in.

Viewer

Media beat-up would have us
subscribe to the white Australia policy
of the turn of last century,
addicted to the screen.
'Detainees' decay behind wire.
This is God's own country?
Action. Interest. Addiction.
We hear and see nothing.

Anchorperson

People-smugglers harness
wealthier refugees and push
them through the pipeline.
What of those left behind?
The Minister for Immigration
asks us to ask ourselves
what right these people
have to jump the queue.
It was rumoured today
that some of these people
have military training.
The gun is part of their body.
Aliens. Cyborgs. Be wary...
Not of us, our clean country.

Viewer

Escaping from Iraq
sanctioned to death,
forgotten war – bombed
with 'our' support;
from the Taliban in Afghanistan,
regime we condemn
when it suits us. The 'elite'
SAS battening down human cargo
rescued from a sinking ship,
watching over our interests.
TV, speak to me.
Tampa Tampa Tampa
destiny.

Anchorperson

They threaten our way of life,
but I retain distance;
they threaten the sanctity
of our neighbourhoods,
but I retain distance.
The people speak, we listen.
The people have spoken, we listened.

Viewer

The Holiday Show, destination
Europe, Asia, the Middle East.
A tele tradition. Diverse range
of culinary delights. Chickpeas,
sesame seeds, tahini, babaganoush.
Modernity gives us access

to all languages. 'Legal' migrants
want 'illegals' kept out.
Prophecy. Pre-destiny.

Anchorperson

... and the gold hidden in teeth
and body cavities is said to affect
reception of cellular phones.
They will form gangs.
They will challenge hegemony.
They will bring law suits
based on high doses of agent orange
from forgotten wars.
They will end up
with representation
in parliament.
Fundamentally.

Viewer

Ocean surrounding us,
we look inwards. The centre
a lung we want to keep clear.
Taxpayers. Apologists
for the Stolen Generation.
Dissemblers of genocide.
The Sydney Olympics,
by jingo!

Anchorperson

Surveys show 101 percent
support for Government policy:
Allons, enfants de la patrie
le jour de gloire est arrivé...
so goes the movie,
jewel in the crown
of the Newest Wave.

–JK

By way of conclusion for now, I just want to mention something about my recent re-entry into Australia. Laden down with baggage, a sick daughter and a five-month-old child, my wife and I asked staff for help with our baggage as we had been advised to do by airline staff. It was evident that we were returning to Australia having lived overseas, having betrayed nation by taking our services and expertise elsewhere, or more bluntly, for leaving the family and its controls. I have time and time again been treated with suspicion in these circumstances, in a way that doesn't happen if you've merely come back from an overseas vacation. Two Qantas hands busied themselves telling us why we didn't need help – were surly, even offensive. The airport was empty by the time we managed to move and they watched on, saying, 'we have to be available to help people who have greater need'. I was so angry, so worn out with well over 30 hours without sleep, with a plane journey marked by a reluctance of Australians to help out, that I said, 'I hate this country'. Things ground to a halt, and every worker within earshot stared at us with contempt. It was a frightening moment.

Of course, everything was then gone through with a fine-toothed comb in Customs. This didn't worry me, I am used to it. The point is, if Australians treat those of us who live elsewhere with such contempt in that quasi-official way, it is barely imaginable what psychological trauma refugees must suffer for their incursion into the family of nation. In these staff, I could see my primary school, and all the other witnessings I've made in my life. It terrifies me. It is a terror casually as well as officially inflicted on 'outsiders' and interlopers on a regular basis.

It is one of the many tyrannies of nation.

CHAPTER 6

Wheatbelt Isohalines & the Making of Isopleths: The 'Annihilation of Distance' & Other Subtexts Associated with the Creation of a Sequence of Poems

It gets saltier around here season by season. The water places have dried out entirely and the white of salt has become whiter. It gleams. It is haunting. If an isohaline is a line that carries us between points of comparable salinity in an ocean, then the pipes that carry fresh water through the Western Australian wheatbelt, crossing salt wastes on small trestles and pylons, are interior or land-bound isohalines.

See Figure 1.

I am often distracted by the nature of the 'line'. Where or what the two points it connects might be in the first place, and the necessary curvature as we plot from one point on the earth to another, no matter how short. Connected at various points to these pipelines are standpipes: the lifelines of the wheatbelt (and elsewhere).

I remember, when I was a kid, going out with my uncle in the truck with large square metal tanks on the back, to the nearest standpipe (maybe ten miles or so away) to collect water for carting back to the farm to refill the rainwater tanks in dry times. The water was for drinking, washing, showering. The toilet was watered from a farm dam, but a month or two into summer would see the house water supplies run dry. For my cousins, these labours were no doubt of minor interest, having to see to them summer after summer, but I always enjoyed accompanying my uncle. I didn't do much. I held the rubber pipe in place as my uncle turned the tap wheel, and the water gushed out. I placed the cover on tentatively before my uncle sealed it shut.

Back at the farmhouse, the water would then have to be pumped into elevated tanks, high up on their stands; the largest stand was fenced in by one of my cousins to house his canaries. Later, as a young bloke doing itinerant work in the country, I didn't find carting water such an interesting thing. Just hard, necessary work.

Last week I was driving down a back road along the Avon River with Tracy and five-year-old Tim, and, as I sometimes do, I stopped the car beneath a standpipe. Tim loves it. He is obsessed with standpipes, these rural life-sustainers ... he fills pages with drawings of them. Ironically, they look something like images of oil wells filling Texas: well after well after well. These don't quite carry that horror, though. Not in intent, anyway. But the images are terrifying. The cumulative impact of the 'mundane': the total occupation of space. A lexicography. A perspective. A flattening or ironing out of depth of field. Even the 'white' between the standpipe images is traumatised by their exponential presence. Colonisation is sequencing. They become close points which are joined by intense, variegated lines. They challenge perspective and functionality: seeing is belonging and belonging is rarely, if ever, linear.

Anyway, as we sat in the car, Tracy remarked on the letterbox next to the standpipe. Strangely, even after years, we'd never discussed how a standpipe functions. She rightly deduced that the letterbox was for recording the water taken: the accounts are kept by shires and it's an honesty system. You note how much you've taken (by reading the meter), and pay for it later. Tim loved the idea. These signifiers on the roadside have letterboxes where the narrative, the story of their watering, is told – and by different voices.

Wherever we drive, Tim counts standpipes – we look out for them. We can map large areas of the wheatbelt by standpipes, and where the pipelines go, where the freshwater flows over the salty wastes. One day, as we drove home from Northam, he asked if we could pull over at the Spencer's Brook standpipe. We did, and I took a few photos. The position of the sun dictated the angle/s. I took one face-on with the line of the pipe aligning with the verticals on the crosshairs of the view. Lines hiding lines. And note the background – all rotates around the fulcrum, the axis of the pipe. Lines of road, fence, instructions, letterbox, trees, are taken into its simple, spatially minute starting point. Trigger of a lexicography. Mind you, you could just as easily drive past and not notice it at all, but gain an impression of the road, the fence, the trees.... But once you know it is there, and once its 'offerings' become life-sustaining, essential, or at least necessary or desirable, it is the centrepiece. It becomes bigger than a big place.

See Figure 2.

Out in salt places, a standpipe ironises and overwhelms the salt, the vista of dead trees or the absence of trees. It is encapsulated, contained ... channelled fresh water. And it is *flow*. *Controlled* flow. I will take a photo of a standpipe out in a salt waste when I get the chance. Ghost/phantom limb ... *not*. Haunting ... likely. It is startling once you appreciate its import. It is such a contradiction that it makes reference to symbolism redundant. It transcends symbolism.

A standpipe is called a 'colonne d'alimentation' in French. Here, among the salt angst of the wheatbelt, there are no sonnets of correspondences. When you try to write them, the correspondences come unstuck. You attempt to write between the languages of imperialism, or you stay trapped in them. The glints out of the salt are fox eyes and/or broken glass. The standpipes are 'pitiless'. We can't begin with Baudelaire:

> La Nature est un temple où de vivants piliers
> Laissent parfois sortir de confuses paroles:
>
> [The living pillars of Nature's temple
> Occasionally let slip muddled remarks. – JK]

I have gone for 'remarks' rather than 'words' or 'speech' as the land is re-marked by each standpipe, and yet they remain so hidden in their openness. People drive by time and time again and don't notice them, if they are not part of their lives. At least this is what I have found with so many of those who have travelled with me. But Tim never misses one. He is sensitised to their presence. He relies on them. The benevolence of availability, the 'posting' of what is owing (recording what is taken, signing off on the deal in the absence of an official), ultimately the 'rights of property' held by the State that the landowner might partake to build his or her own profit, that it might feed back to the State and increase its strength.

Last winter I watched a shire truck with a huge spray-tank on the back using the standpipe going out of town towards Quairading to embody the poisonous swill – to mix herbicide with the necessary quantity of water. The standpipe's rubber hose was deep inside the spray tank. The next users unwittingly dip the same hose into their tank for fresh household drinking water. Not that many would see it as a problem. It's a spray-crazy shire. Standpipes are also flagpoles. From the taps flowing with scheme water made potable for us all ... no matter how isolated, we are 'assimilated'. That's terror.

I have been trying to write 'standpipe poems' since I was sixteen. Somehow, although they occur as isolated references in poems, they have never 'come off' for me as satisfactory entities in themselves. Rather, a sequence of fragments has been accumulating over the decades. Abandoned poems are to be salvaged (or in the case of the violent world I come out of, maybe 'gutted' is more disturbingly appropriate); links and new segments to be introduced: segued in, interpolated....

> salmon gums' lessening
> intensities, signposting
> where family flourishes

like the standpipe
at Wickepin, eternal source
of water, nose-turning
at drought, like death
everything is perfect
and decomposition
a sonorous speech,
first rains and the pipeline
deflecting silver-skinned
accentuated song
an outsider would romanticise
and take back to a place
of denser population

(From an abandoned, unpublished poem, 1999, 'Below, the narrative'.)

So 'everything that'll carry water'
stacked on the tray-top,

blue Ford truck gathering speed on gravel,
low to high, pulling in under the standpipe.

Through the canvas hose
water rushes like there's only water,

and the dry about is travesty,
a bad joke. It is cooler than water

should be. It is literate.
It burns and soothes at once.

To pump back up into the house tanks,
to force and coddle gravity,

to cosy up to the flow before the boss
claps it shut, says waste not –

it's a long haul back to the standpipe,
and winter'll be slow in coming.

(From an abandoned, unpublished poem, 2004, 'Standpipe'.)

The time signature is variable. From 'spots in time' through to compressions and elisions of time. Memory is made active, memory is twisted. I go out to a standpipe with my young son and we stand beneath it. He enjoys the tension of it possibly – with no logic – coming on and us being drenched in cool, clear water out near the salt wastes and dusty paddocks. The heat is intense – we can't be out long in the open. The risk of sunburn is high. Of skin cancer. I've already had a bunch of skin cancers removed, and though hatted and well covered, his exposed five-year-old skin is especially vulnerable.

I believe poetry is action, and poetry emerges from experientiality, no matter how separated from the moment. I visualise these lines:

> Relativity of coils,
> serifs of telegraph wire
> offsetting railway lines
> burnished with heavy
> harvest, rolling-stock
> steady all hours of night
> though further out
> dust even in winter
> and crops dead
> before they emerge,
> beneath the bend, the crook near the road, an arrested *paraphe*,
> underscore of the thirsty, transliterating a signing off:
> water come up from catchment where less rain is falling,
> and gravity still regulates.

Old telegraph poles used by the railways, standpipes, roadways, they all muzzle through the narrow 'public space'. The links between the lines, or as McKenzie Wark might call them, 'vectors' (see for example Ken's *A Hacker Manifesto*), are increasingly fascinating to me, as they are to others who have lived in Australia. I do a Net search. I am impelled, though I'd rather not interfere with my thought processes. I find a radio station site – 2SR-FM – and this, under the title of 'Australian Wire Music', posted by 'Warren' on 25 May 2005. It will open doors down the track, no doubt:

> Post-contact outback Australia is a land that has been divvied up and delineated by wires – wires nailed to fence-posts to mark out properties; wires strung atop poles to deliver power and telecommunications. If you've spent any time driving through it, these wires (and the road under your wheels) are often the only discernible reminders of the European conquest of this sparsely populated continent. (Until, of course, you come across the inevitable roadhouse/internet café in the middle of nowhere.)
>
> So, it's somehow appropriate that the two artists who have been most active in exploiting the musical potential of such long-span wires, are both Australian.
>
> First up is Alan Lamb, who records the sounds made by telegraph wires as they vibrate in the wind. He started doing this in the late 80s, after acquiring a property near Fitzgerald National Park in southern Western Australia, that was traversed by these wires. The

> recordings he made were done by attaching contact mikes to the wires and waiting for the wind to set them in motion....
>
> Unlike Lamb, who passively records the sounds produced by wires, avant-garde violinist Jon Rose seeks out continent-spanning barriers like the rabbit-proof fence, and hammers and saws tones out of them with help of violin and cello bows....
>
> <http://www.filmcement.org/rummage/archives/000193.html>

This is exciting. I will need to track down the music and no doubt a segment of the sequence which is now moving from standpipes to telegraphs and other wires (I have written dozens of fence-wire poems over the years), and other people's interpretations mediated through my own might come into play. I make a note to go back to this when I get a moment.

See Figure 3.

I have been writing a telegraph poem based on a reading of Stendhal's *Lucien Leuwen* (more of which shortly). Here's some of what I have already:

> And so, my brother and I were taken places
> where telegraphs had died out, thin bodies
> debased by salinity, burnt back to carbon dating,
> hacked down for fenceposts.
>
> We collected
> ceramic and glass insulators, sometimes strands
> of the peculiar wire that carried messages,
> carried the long and the short of it,
> wire that wouldn't coil easily,
> kept springing out like a slinky,
> snake-like. We made our own fun,
> adults would say.

Other chunks of these 'notes' to a possible poem include the following fragment:

> Telegraph remnants bring anxiety
> and an urge to collect: pioneer
> incendiary aloneness. Recollect
> taps, drum taps, bloodtrail
> of signal and severed tell-tale,
> gossip of the give-away.
> Birth and loss, an echo.

There are also notes about Paris, Louis Armstrong, and Freud on Leonardo Da Vinci's childhood. Eventually, I am sure, out of these diverse fragments, a sequence will form. I want it to be unified in form, but not in intent. I want it to be 'subversive', and 'undoing'.

But for now, I have another issue. Morse code. Morse code stopped being used when machines started interpreting messages. Then it became Baudot code or ASCII. I have been working in Five Unit Telegraph Code. Poems written in code – ITA2 code. Plenty of poems have been written about telegraphs but when was the first poem transmitted by telegraph? I have been researching this. I contact Richard Bright, Hon. Secretary of the Morsecodians Fraternity of W.A. Inc. I ask him numerous questions about the history of the telegraph in Western Australia which will feed into the sequence. I mention this essay, an essay on process, not end results. Here's a snippet of what he said to me in a private email on 1 January 2008:

> I think the only place that may hold maps of old telegraph routes would be the Battye Library [in Perth, where State historical records are held] – I did have a list (1935) of PMG [Post-Master General] telegraph lines in W.A. – I will see if I can locate it and let you have a copy.
>
> I know the PMG Dept did use old railway lines as poles, but so did the WAGR [Western Australian Government Railways] in their telegraph network. I think the one you mentioned between Northam and York would probably have been a railway circuit. The WAGR used the Electro-Magnetic Telegraph as well as the PMG, however their network was not nearly as extensive and was for their own private use i.e. it was not a public message system. The WAGR lines followed railway tracks and only connected major railway junctions/terminals i.e. Perth–Northam–Merredin–Kalgoorlie. They also had telegraphists stationed at Katanning, Albany and Bunbury and maybe others – I'm not sure.
>
> The first telegram transmitted in Western Australia was on a twelve-mile long 'wire' between Perth and Fremantle on 21 June 1869. It read:
>
> TO THE CHAIRMAN OF THE FREMANTLE TOWN TRUST. HIS EXCELLENCY COLONEL BRUCE HEARTILY CONGRATULATES THE INHABITANTS OF FREMANTLE ON THE ANNIHILATION OF DISTANCE BETWEEN THE PORT AND THE CAPITAL AND HE REQUESTS THAT THIS FIRST MESSAGE MAY BE PUBLICLY KNOWN. GOVERNMENT HOUSE 21ST JUNE 1869.
>
> <http://members.iinet.net.au/~oseagram/wahist.html>

The issue of occupation and space obsesses Australian literary theorists and commentators as much as geographers. Vast distances are a source of national pride concerning 'bigness' – consider Les Murray's brilliantly epiphanic and anthemic poem 'The Quality of Sprawl' (in his *The People's Otherworld*) – and yet the bridging of these gaps is inseparable from the creation of nation. Last year ABC Television ran a series of nationalistic and propagandistic programmes on great engineering feats of nation, for example. One of these, entitled *A Wire Through the Heart*, documented the establishment of a telegraph between the north and south of the continent. As explained at the Film Australia digital learning site (for schooling):

> In 1870 Charles Todd, using explorer John McDouall Stuart's maps, organised and led three teams to lay the overland telegraph wire. In eighteen months Adelaide would be linked to London via the undersea cable to Java and then on through India and so Australia was connected to the world.
>
> The international telegraph link from Java would come ashore at Port Essington, in what is now the Northern Territory. Stuart had provided a route for an overland telegraph line across the forbidding interior of the continent. All that remained was to build a telegraph wire from coast to coast, passing through the arid heart of Australia.
>
> <http://dl.filmaust.com.au/module/1013/>

The paradoxical 'arid heart' is such a core image in nationalistic Australian thinking. The spirit of Australia is to be foreboding. This centres (literally) on 'belonging'/occupation, and the exclusion of Indigenous peoples, of claiming aridity as nothingness (to delete their presence), but to 'irrigate' that aridity with modernity and the new (colonies, later nation), through technology. The words that pass through the heart become its 'real' singing, as opposed to the unseeable (if you don't look or listen, or don't know how to look and listen) songlines of Aboriginal people.

By 1881, the telegraph line went from Perth, the capital of the colony of Western Australia, to Eucla, on the border with the colony of South Australia. Western Australia was connected to the world. One of the landmarks noted by the Western Australian Fraternity of Morsecodians is '1929. The gothic typewriter, namely the Imperial model 55 typewriter, was introduced.' I have drafted poems on a typewriter like this; I have also discussed a 'wheatbelt gothic' in my poems, and have a poem that goes by the title of 'Wheatbelt Gothic'. A synchronicity that seems hard to miss in terms of writing a poem on telegraphs, a sequence of poems. As decoration, illustration, compulsion, or because it completes a picture? All of the above.

The telegraph evoked literary writing from its earliest manifestations. One of the great works of fiction 'about' the telegraph is Stendhal's deconstructive *Lucien Leuwen*. Poets were in there too, writing odes and celebrations to the telegraph.

All of which brings me to my point about the isopleth, telegraphs and environment. An isopleth, to quote the *Dictionary of Geography*, may be defined as 'A line on a map drawn through places having the same value of a certain element. Such lines show the geographical distribution of the elements. Familiar examples of isopleths are isobars, isohyets, isotherms'. The isopleths I am identifying are lines on a conceptual and literal map connecting places of environmental degradation (massings of lines in all directions) and those of environmental 'preservation' (few, but dramatic). In this poem-essay, which will follow the construction of a sequence of poems, I will follow telegraph lines: present, removed, and/or lost. They are the sub-narrative of the poem.

I wrote once, in a sonnet about sonnets, that Edwin Denby gave insulators from the top of power poles to his friends. I related to that because, as a child, I too collected such insulators. The insulators I valued most came from telegraph poles, which still criss-crossed the wheatbelt but were even by the early 1970s abandoned and dysfunctional. I was fascinated by these insulators not only as things in themselves, but as markers of what had been, and points of separation between the message – the Morse code speeding through the fence wire stretched between wooden poles, often made from local woods ripped out of paddocks being cleared – and the earth itself.

The wooden poles were like ironic comments on intermediaries, negotiations between the human-made and the natural world. The telegraph almost grew out of the ground, but it signified the end of nature. It also signified, as is well attested, the basic colonial urge – and its vulnerability. A line could be so easily cut. And its markings and stretchings formed grid-patterns and stringing together of ownerships. The telegraph systems that belonged to railway systems, and these posts, as indicated above, were often made from railway lines with wooden cross-beams to insulate, especially against lightning strike. I grew up knowing the man who introduced wooden cross-beams into Malaysia as protection against lightning strike. Lightning and telegraphs, as much as standpipes and fences and gravel roads, bind my wheatbelt world together.

I am driven by a necessity to write, and I often work with recurring motifs. Sometimes the compulsion overwhelms, and even if a poem is not 'there', I take notes towards what will one day become a 'poem'. Often long-term fascinations and compulsions form into sequences.

I was recently working on something for a French literary journal on the nature of drafting. The fact that it was a French literary journal is pertinent for a number of reasons. I sent the journal photos of where I'd been working on the drafts of a poem they were running (as drafts), and in a number of photos, next to a manual typewriter I was using in a work shed, was the top of an old telegraph pole, complete with insulator. In reading and doing some work on the 'unfinished' *Lucien Leuwen*, one half of which is entitled *The Telegraph*, I have been doing a lot of thinking about the interconnections between the telegraph and the Internet, and how these might pertain to poetry. Many critics have also drawn parallels between telegraph and Internet, and Stendhal's work certainly gives you a sense of networking, a new sense of time, the possibility of interception, and a new political landscape being formed by such networks of communication.

I have also been translating Paul Verlaine recently, and his wonderful poem of travelling/moving (in a train) also adds to my purpose in creating a parallel movement to the unseen movement of 'speech' across the wires. I travel vast distances through the wheatbelt by car, as I record environmental damage and renewal, constantly aware that my drives contribute to the undoing of the environment in so many ways. The balance of irony weighs heavily. It builds subtexts to the poetic texts I am creating. Paul Verlaine's poem 'Le paysage dans le cadre des portières' (from the collection *La bonne chanson*) brings to mind Ross Chambers' essay 'Training for Modernity'... and so the network of in-sourcing builds. First, the poem:

Le paysage dans le cadre des portières

Le paysage dans le cadre des portières
Court furieusement, et des plaines entières
Avec de l'eau, des blés, des arbres et du ciel
Vont s'engouffrant parmi le tourbillon cruel
Où tombent les poteaux minces du télégraphe
Dont les fils ont l'allure étrange d'un paraphe.

Une odeur de charbon qui brûle et d'eau qui bout,
Tout le bruit que feraient mille chaînes au bout
Desquelles hurleraient mille géants qu'on fouette;
Et tout à coup des cris prolongés de chouette.
– Que me fait tout cela, puisque j'ai dans les yeux
La blanche vision qui fait mon coeur joyeux,
Puisque la douce voix pour moi murmure encore,
Puisque le Nom si beau, si noble et si sonore
Se mêle, pur pivot de tout ce tournoiement,
Au rythme du wagon brutal, suavement.

In my translation:

The countryside in the frame of the carriage doors
Rushes wildly by, and entire plains with water,
Wheat, trees and sky are engulfed
More and more by the cruel whirlwind
Which swallows the lean poles of the telegraph
Whose wires look strangely like a paraphe.

A smell of burning coal and steam,
All the din that would be created by the scream
Of a thousand giants whipped on the end of a thousand chains;
And suddenly, the owl's drawn-out cries.
– What is all that to me, since in my eyes
Is the white vision that makes my heart joyous,

> Since the gentle voice still murmurs for me yet,
> Since the Name so beautiful, so noble and so resonant
> Blends suavely, pure axis of all this whirling,
> With the rhythm of the brutal carriage.

Ross Chambers, in an occasionally brilliant essay on Verlaine and the process of reading a poem, captures the essence of this poem when he writes: 'the metaphoric transformation of the telegraph-lines into a "paraphe" (a certain form of writing), and also ... the contrast between the "suavité" of his mental images and the "rhythme du wagon brutal"' ('Training for Modernity', 162). He continues:

> The 'telegraph' is the industrialized form of writing and communication that corresponds to the industrialized mode of travel that is the railroad. The two inventions, historically speaking, implied and supported each other; and the reason the telegraph-lines ran parallel to the track was that the railway could not operate without a signalling system. Is the poem perhaps describing itself as *running parallel* to the railway experience it is describing and part of the same system, rather than directly modelling itself on the train? Notice too that the 'tele-graph' is long-distance writing.... (173)

It strikes me that this would have been a far more interesting observation had it been made in the era of the Net, which in 1990 really hadn't gained a foothold.

I have been writing a series of 'graphology poems' since the mid 1990s and now I realise this sequence I am envisaging is also part of the greater sequence of graphology poems: poems about writing, recording, inscription, observation, and so on. I am up to the 700s with those poems (irregularly numbered, with gaps in the numbering), so this standpipe–telegraph–wire sequence will be 740 onwards....

And so Ross Chambers progresses with a reading of Verlaine's poem. What he is theorising, I am trying to put into action. In an age of 'lit crit' and 'theory', the absorption of critiques into the poem is commonplace, and even critiquing those critiques in the poem is to be expected. But for me, reading the poem in the landscape I am part of, Chambers' critique and the original poem are politically problematised or complicated. He refers to the relationship between syntax and what is said regarding the telegraph posts 'swirling into the same all-engulfing abyss' (173). In a land like Australia, where Indigenous inhabitants had their land stolen in such a recent and direct way, the implications of abyss must be investigated. The reading of a Verlaine poem, and the reading of a linguistic critique of a Verlaine poem, are perverted.

See Figure 4.

Of course, for me reading in situ, out here, the whirlwind is the dust devil I photographed a couple of days ago. The wheat, trees (many dead in saline scalds), and sky, are there in significant proportions. The engulfment, the 'abyss', is coloniser dross. It is imperialism of space and spatiality. It is the 'annihilation of distance' of the telegraph, the filling-in of the 'emptiness' (and ignoring of the marks of previous and simultaneous occupation.... I no longer think in terms of palimpsest and overlay of the Indigenous, but rather of the blindness and refusal to recognise the signs which, as the artist Julie Dowling once told me via email, can never be deleted or overlaid).

Chambers discusses the belated appearance of *'je'* in the poem, but the issue of subjectivity (and delayed subjectivity) has different possibilities out in the induced salt-wastages of the once scrub-rich and animal-rich (certainly relatively) spaces of the wheatbelt telegraph.

I have long wanted to write (and have written) poems in Morse code. I want to find a remnant network and send them – though one can download software and send them in this form across a computer network. If you've ever heard the wind working old telegraph wires, it is not surprising that a Western Australian composer, as mentioned above, has used them to compose music. He talks of the 'ping' recorded when insects strike unsheathed wires. I can relate to that. As with standpipes, the repetitions of observation and familiarity, mixed with reading and comparative interpretation, build an unwritten sequence.

In essence, I could take the poems I have written 'on' telegraphs over the last thirty years and *make* a sequence: that is, a simple combination of different poems with a common theme, linked by a genome, linked by the code that Chambers mentions or that Wark denotes in his conceptualisation of vectors. That might be enough. And this would be a life's project, but not enough for my purposes. I want to cast back, but also to open new space (in a non-colonising way!) and project a way forward. The first telegraph poem I wrote and still have a copy of was written when I was a teenager. It is straightforward in its expression. It simplifies the encodings:

Passing the Ruins

The bus driver had said
that the old telegraph station
now covered in drifting sand
was operated by telegraph men
from the two different states

and that they sat on opposite
sides of a partition passing
messages blindly through holes.

The key is 'different states', and it's the unseen connections between different realities that still interests me. The sequence I am now trying to write comes out of this earliest thought. The person sending the message, and the person receiving the message, have (often) little thought for the space it traverses, the damage their message outrides, like fresh water through eroded and degraded areas, travelling in its insulated piping.

The image I have to connect to this conceptualisation is of the semi-famous Pink Lake outside Quairading, on the road to Bruce Rock in the Western Australian wheatbelt. Pink Lake is heterochromic: often pink on one side and, across the road, blue-green on the other. It is much photographed. It is cited as an example of 'dryland salinity', because the massive clearing around it has led to concentrations of salt and the death of the trees and scrub that once filled it. The wheat crops surrounding this eye of hell glow and yield their ironies and wealth. It's a short-term fix. Not far down the road, the stubble residue of wheat crops has been grazed to dust by sheep, and a dust devil sucks the dirt into its funnel. It yields and yet verges on ruin through drought, or salination through rising water tables. That's the irony of the wealth extracted from this landscape.

But to the telegraph: what fixes me to Pink Lake is the line of telegraph posts eaten away by the salt (topknot pigeon sitting on an old telegraph post, lacunaed by salt clusters), like the fence posts, those demarcation lines of ownership and denial (of passage), in such visually arresting ways. The defeat of the colonial symbol of the telegraph, empire-builder, is the symbol and machine of the damage done. It consumes itself, that old Pacman image. Either side of the lake, the telegraph posts have basically been removed, so they are preserved until they are eaten away by salt as a memorial, or maybe a memento mori. Those missing posts (whether treated in creosote or not) have no doubt been burnt in farmhouse fires or in bonfires during clearing episodes. That top of a telegraph post I have is a rescue item from a pile of sawn household firewood.

The sequence. Where do we start? An image, some reading, a history. A sense of place. A desire to preserve. Reverence and damnation. The vital thing for me to understand, as well as my readers (not that it's hard to see!), is that I am compelled to write not because I adore 'nature' and want to be in tune with it, but because I am outraged and angry, because I feel, seriously, every bit of damage done. I drive past a pile of scrub from a piece of freshly cleared land, and I become sick and depressed. I loathe the spectacle of usage, and can't see logic in it.

The other day I was driving a guest through the forests, and objected to a communications tower. She was bemused and said, 'But people need to communicate'. My answer: not that way. Those who know me would be bemused to hear of one so plugged into the Internet saying such things. And it's true, I was using and building and programming computers back in the late 1970s as a school kid. But I am looking to move away from it all: I am working to dismantle. Not by viruses of computer code; by viruses of poems. Here's McKenzie Wark, the cultural critic who has written extensively on 'vectors' and communications, including telegraphs in literature (and elsewhere):

> In epidemiology, a vector is the particular means by which a given pathogen travels from one population to another. Water is a vector for cholera, bodily fluids for HIV. By extension, a vector may be any means by which information moves. Telegraph, telephone, television, telecommunications: these terms name not just particular vectors, but a general abstract capacity that they bring into the world and expand. All are forms of telesthesia, or perception at a distance. A given media vector has certain fixed properties of speed, bandwidth, scope and scale, but may be deployed anywhere, at least in principle. The uneven development of the vector is political and economic, not technical. (*A Hacker Manifesto*, 23)

I refuse to use mobile phones. That's one of my lines in the sand.

I write a sequence of poems that is encoded in such a way that it will spread. It is a worm.

Guy Debord opens *Society of the Spectacle* with this aphorism:

> In societies where modern conditions of production prevail, all of life presents itself as an immense accumulation of *spectacles*. Everything that was directly lived has moved away into a representation. (Thesis 1)

I seek to sabotage this representation. Is this an action compatible with my vegan anarchist pacifist politics? I think so: it's an activist poetry. I find that much poetry, including that which paratactically pretends to be dissolving and disconnecting, still works towards a compliant unity of receptivity to this representation. Unless we change our lives, the poems we write become more lies. This is the second part of Debord's 'manifesto':

> The images detached from every aspect of life fuse in a common stream in which the unity of this life can no longer be reestablished. Reality considered *partially* unfolds, in its own general unity, as a pseudo-world *apart*, an object of mere contemplation. The specialisation of images of the world is completed in the world

> of the autonomous image, where the liar has lied to himself. The spectacle in general, as the concrete inversion of life, is the autonomous movement of the non-living. (Thesis 2)

It sounds like all the pre-requisites to making a poem. And this is what I am doing as I am writing and lying to myself. Lying? I am telling it as I see it, as I experience it, and as I feel it. But it is 'participation' in the spectacle, making a pseudo-world apart. Yet I need to do this to corrupt the unity of the reader's vision of security. Nature writers who love and lament damage to the natural world, but still damage it in so many ways, still maintain the footprint they protest against; they can make beautiful poems, but the anthropomorphisms can't be got around.

Here's a true story. My brother, whose art is much talked about by those who know it, refuses to participate in the art world. I extracted some images from him for covers and things I've done recently, but that's it. He is truly hermetic. He despises the world of art and literature. He wrote poems once, but destroyed them all. He believes his 'nature' poetry is only useful to those who would read it as useful: compliant to their already satisfied view of nature. I see his point.

Let's skip to Debord Thesis 4: 'The spectacle is not a collection of images, but a social relation among people, mediated by images'. The sequence of poems, the sequence in a single poem. The desire to speak to other humans, not to nature: to tell them, cajole them, convince them, advertise for a like-minded partner or reading circle.... Images are the softeners or the head fucks. Like the cult groups using mind-control and sex.

I am writing the sequence now. It is in alternating long and short lines. There's a drive involved. I am going from York to Quairading, following the train line and the old telegraph lines. Then out of Quairading to the Pink Lake and the poles being eaten; out further to the Yoting wheat bin, left turn on towards Mount Stirling. Massive clearing here. The lower levels are just salt wastes. I stop to photograph old telegraph posts along the roadside. They start beheaded, and become whole nearer the great granite rock of Mount Stirling, named after the first Governor of the Swan River Colony. It is a nature reserve now, though under stress from surrounding farms, tourists, and 'introduced animals' such as foxes and cats. That's another issue.

The telegraph poles have one or two insulators, but in the main, they've been 'plucked' or 'harvested'. People – and it's not only *kids* – collect them. Near Mount Stirling, I see a magnificent bungarra on the road side. The bungarra is a monitor goanna, very closely related to the racehorse goanna. They are so hard to tell apart that there is much debate on where one begins and ends, and *how* you tell them apart. Nomenclature. Epistemology. Lexicography. Secodont: cutting words. Create neologisms. The *paraphe* of Verlaine. I turn back to photograph the monitor. As I prepare to frame the photo, a car passes, swerves, and strikes the great majestic reptile. My daughter had just remarked how regal it looked (four feet long, up on its front legs, long heavily-skinned neck)....

I told my daughter to look away. It was a lump of dead flesh. Humanity. That's it. You have to try hard not to hate. The annihilation of distance as we drive, as we listen, as we write and send our message poems, is literal. Even vicariously. I didn't kill the animal, but I also drive a car. I have accidentally struck animals with cars. Tawny frogmouths at night, mice, frogs; side-swiped kangaroos, swerving hard to try to miss them at high speed. I have lamented these happenings over the phone to my brother. The poem's lines are concentrating. Break the sequence. Next part. A possible pattern is forming. Part 3. I think of Verlaine's poem. Part 3 will be a translation of his poem. Part 4 will be a variation on that poem. I am creating Isopleths between poem sections and landscape, between events in my life and events I acquire from outside my life. The nexus points of reading and experience. The sequence is a series of Isopleths, of crisscrossing telegraph wires boosted by prosody, the repeater stations that drove the telegraphs. In essence, the images are the seeings: I see the poems in my head as I see the composition of a photo before I take it. I storyboard the poem. The sequence of poems is embedded in the script of this essay: the golden rule of paring back is triggered, and vast swathes of words disappear into Morse....

Spectacular Pink Lake
heterochromic
monitor gothic scripts
heat and topknot pigeon
perched on the plucked
dead upright, regal,
struck down.

..-. .-.. .. --. .... -   -.. .. ... - .- -. -.-. . ...
.-.. --- --- .--. ...
-.. . .- -.. .-. . -.-. -.- --- -. .. -. --.

The process distils. The Isopleths are becoming clearer and the sequence within a sequence is almost ready to be mapped. The code will shape lines: short and long, the alexandrines of Verlaine, glimpses in large spaces. The lines I draw are like the lines of the telegraph and the pipelines that find the standpipes, but they are not the same. They are conjectural lines drawing conversation. They do not seek to annihilate distance, but to celebrate it. Their anger is in the failure of intent behind the laying of those lines of communication: to draw close together by ignoring what's in between.

CHAPTER 7

Half-Masts: A Prosody of Telecommunications

> ... we no longer have roots; we have aerials.
> – McKenzie Wark, *Virtual Geography*

For some years, I divided my time between the city and the country. When I was in the city, I often used to spend time at Wireless Hill near the Swan River in the southern suburbs of Perth. It's now a recreation and nature reserve. I found this an interesting place because of the presence of native bushland in an area that had been otherwise entirely cleared, and filled with housing, but also because of its history. The Friends of Wireless Hill have noted the following on their website:

> Wireless Hill was once known as 'Yagan's Lookout', providing perfect views of the surrounding area. Yagan was born in 1810 and was the son of Midgegooroo, the leader of the Beeliar tribe who were custodians of the Melville, Fremantle and Cockburn districts. Yagan was a well known figure in the early days of the Swan River colony, respected by the settlers for his strong personality and independence. He also advocated peace, believing blacks and whites could live in harmony.
>
> In 1912 the facilities at Wireless Hill enabled wireless communication to be carried out for the very first time between the east and west coasts of Australia, between the mainland of Australia and shipping up to 1,600 km into the Indian and Southern Oceans, and between Australia and the rest of the world.
>
> <http://www.sercul.org.au/wirelesshill.htm>

In this simplistic and hebephrenic account of 'Yagan's Lookout' – Yagan was murdered and beheaded by white youths [see Coughran and Laurent, 'Please Take Me Home' – NL] – we see a typical absorption of Indigenous presence through assimilation into a white timeline. Wireless Hill, as we find in the second paragraph above, was an extremely significant telecommunications site through to the 1960s. Not only did it allow for the first contact between east and west, but it was also a key point in military communications for that period; so the suburban enclave became a radiating centre of State. The most distinctive feature of the landscape was the three massive concrete towers, on top of the hill, which had anchored the giant aerial. During my early visits, all that was present apart from these towers was a small telecommunications museum and a moderate-size aerial used by amateur wireless operators. A more recent visit brought the shock of a huge mobile telephone tower emanating high amounts of electromagnetic radiation. Next to this there is a children's playground and a picnic area.

The extract from the Friends of Wireless Hill website makes (I suspect) an unconscious colonisation of indigeneity into the history of the occupation of airspace with radio and other waves. Not only has the land been taken – its earlier pre-colonial status strangely emphasised by the fact of the remaining bushland as an island among the larger visual overlay of suburbia – all around, and you can certainly see a long way from up on Wireless Hill, space is occupied vertically as well as horizontally by radio transmissions to and from the telecommunications mast.

Radiation is rhythmic. Telecommunications are textual. The combination makes for a prosody, a theory of versification that is colonising and in the end destructive. This destruction, too, is rhythmic. Tetrawatch in the UK, a research-based campaign against the detrimental effects of Tetra mast radiation, tells us on its website that

> TETRA masts pulse at 70.56 Hz (4 × 17.64). However, in this pattern the first and last pulses are not separated, so there is a steady rhythm 'di-di-dah-di-di-dah', 17.64 times a second. The music of TETRA? Our bodies recognise patterns and rhythms and frequencies exceptionally well: this is not raw energy, it is information to us, interfering with our own informational bioelectromagnetic systems.
>
> <http://www.tetrawatch.net/tetra/pulse.php>

Of course, as poets we will recognise the di-di-dah as an anapaest, and will immediately interpret these rhythmic possibilities as forms of poems. The hearing of a poem, as we know, operates on many levels: the pleasure of the experience of hearing does not mean that we haven't undergone some kind of psychological shift that we are unaware of, and which could be positive or negative. So it is with the radiation of telecommunications. Ironically, as we stand within the vicinity of mobile base stations, or indeed use mobile phone technology (which I don't), we are converting through speech and text innumerable 'sound poems' which we can't see but which can be registered scientifically as pulses. So our words, whatever they might be, also carry a different kind of sound we don't necessarily hear.

When I spent time in my early years at Wireless Hill it was often with my young first son on access visits, having been separated from his mother. I went there because of the mixture of bush, which I love, and general recreational facilities. I wouldn't do that now. Unsurprisingly, when I wrote poems out of these visits, I wrote not only of the disjunction between the natural environment, the history of telecommunications, and the realm of suburbia that surrounded it all, but also of a history of the unseen yet prosodic waves that had filled the atmosphere around this area and echoed out. One such poem is entitled 'Wireless Hill' (see my *Poems 1980–1994*), and the relevant stanzas concerning the unseen textuality of conversation are as follows:

> And from
> the central tower I look out over
>
> Alfred Cove, and absorb the river.
> You watch the children ski down slides
> in the adventure playground and scoot
>
> their bikes about the walkways, the sticky
> hum that comes with rubber on hot concrete
> reminding you of our son. You look
>
> to the base of the tower, I look out –
> even further than the river. But the sun
> drives us toward the shade and touching
>
> earth we hear the silent conversations
> that crackle so faintly, too faint even
> for aerials to detect.

Where I live with my current partner and children now, at the gateway to the central wheatbelt in the Avon Valley of Western Australia, we are surrounded by low-lying, eroded hills that were once the core of the Dyott mountain range. Directly behind our house is Walwalinj or Mount Bakewell, the highest point in the wheatbelt. Throughout the wheatbelt, the highest points are crowned by telecommunications masts. Being the highest point, Walwalinj is particularly encumbered with such apparatuses. Stories in town go back thirty years about old-timers wanting to fix them with a stick of gelignite. That was when the main tower primarily looked after television, radio and police communications. There is also an aircraft beacon up there, used by the military. In the last few years, mobile phone and wireless technology base stations have been added.

See Figure 5.

Our son Tim looks out of his bedroom window onto the beauty of the mountain, and its corruption in the form of telecommunications towers.

As with most high places in the wheatbelt, it was where explorers, in this case Ensign Dale, surveyed the land below and in essence laid claim to it for the Crown. In the same way that Yagan's Lookout becomes an historic euphemism for the attempt to 'blind' the traditional owners of the land, so too does Walwalinj, or Mount Bakewell, as Ensign Dale named it, establish a point of surveillance from a sacred Indigenous place. The telecommunications towers are the acme of that surveillance and occupation.

See Figure 6.

The phantom egalitarianism of telephonic exchange: the ordinary citizen sends messages to the tower as much as receives them, and private industry has staked its airspace, its radiation space, though in fact it is still mediated and controlled through the rules and regulations of the State. Its claims are vertical and horizontal. Its claims are the shape of poems. In order to placate then four-year-old Tim, whose fear of these masts is almost fantastical, we offered to drive him round the back of the mountain so he could see them from the other side, because he thought they were monsters. At the time I was working on the *Inferno* canticle of my 'distraction', Dante's *Divine Comedy*, and it was inevitable that this experience would work its way into the text. Here is the end result, part of my own *Divine Comedy*:

Anti-clockwise Canto of the Moebius Strip: back of Walwalinj – 'the mountain' (Inferno 34)

So Tim has seen the back of the mountain,
asking us to turn back home in case we end up
any closer to the communications masts

than we already live. We drive through
the passageway where Satan-legs wave
like tendrils of cloud, and farm dams

look so suspicious that Tracy and I
spell things out: 'have you noticed how
much dead oats swirled about around

the mouth of farm dams look like p-u-b-i-c
hair...'. Be careful, he's learning
to spell fast. Tim's fear of the back

of beyond might have something ...
an extra antenna is visible, having gone
up unseen over the last year.

On Walwalinj, with its powerful spirit,
a devastating energy whips and lassoes.
At home, Tim draws roads snaking

about mountain on which red masts
bristle. He draws them from every angle:
an exact science, a diorama
  of love and loss.

There are only fragments of the original bushland left in the Western Australian wheatbelt. Many of the reserves that remain are small, and are ultimately collapsing ecologies because of the lack of corridors connecting them to other remnants. Quite a number of these small reserves are on comparatively high ground, targeted by telecommunications personnel as ideal locations for their masts or towers, and so are inevitably damaged, not only by the radiation from these structures but also through the sheer act of installing and maintaining the equipment.

We live about 800 metres or perhaps a kilometre from these masts. When the wireless technology was installed, I wrote to the company complaining about the risk they they posed to my family, especially my children. These were the questions I put to them:

1. did you clear or damage any bush on mnt bakewell installing your tower?
2. what assurances can you give locals re health issues pertaining to emissions from the tower?
3. why wasn't there public consultation about its installation?

Their reply was polite but said that their installation complied with national standards and was completely safe. I challenged that with a series of data from investigations done in the USA and the UK, which clearly showed evidence for their safety was in the least inconclusive. This time I received no reply.

In writing this place, its ecology and its histories, which are tormented and sanitised in the way they are conveyed, not only to visitors but to many of the people who live here, I am constantly confronted with the visual aspect of these telecommunications masts, and the knowledge that their language, their poison, is written all around me unseen. One can write about wandoo trees, or marri crown decline on Walwalinj, but not to mention the communications masts would be an error of both fact and aesthetics. Bizarrely, there exists a practice in some places of disguising mobile phone towers by either making them look like vegetation, specifically trees and palms, or even masking them entirely in these forms.

I have long been protesting not only against mobile phone base stations, whether they are in country areas or clustered next to dense housing, or hospitals, or hotels, in the city, but also against the possible health risks of the actual phones themselves. More recently, I have extended this to wireless Internet technology with its deplorable colonisation of public and private spaces, with so-called 'hot spots'. This is language that bends itself without much difficulty to poetry, but references alone to potential problems with this technology do not make a poem an activist poem. What does?

In travelling through the wheatbelt and mapping the lines of connection between telecommunications masts, one is writing a map not only of technology but also of 'uncommunication'. Recently my mother asked a telecom representative about the safety of birds and other creatures around the telecommunications mast up on Walwalinj. The short, sharp reply came: 'Well, if the birds perch on some of that communication equipment, they're fried.' Whether this is truth or exaggeration, much evidence exists that wildlife suffers as well as people. These artificial 'trees', these potential roosting and perching places, become zones of death. Gaston Bachelard, in *Air and Dreams*, asks:

> What is the bird doing on the great stone tree? Is he not adding his wing to motionless height? Rigid treetops are not entirely aerial. The dynamic imagination wants everything in the heights to move. (210)

If the imagination compels us to make the stilled move, if we are compelled to communicate, there is an unresolvable contradiction in these 'trees' whose movement is damage. The story of Babel, in which the tongues of the workers were confused and no single language was the same, was a result of hubris. Discussing mobile mast towers in this light easily becomes humorous or bathetic, because we don't usually see the consequences; because there is no immediacy, there is irony.

There is an entire poetics of trees as both vertical and horizontal entities. The vertical is obvious, but the horizontal is visually defined in branch structures, and also in the unseen root structures. In a sense these telecommunications masts and towers function as simulacra of trees on high places (one might think of the 'high places' of the Hebrew scriptures), as altars of exchange. These steel structures barely move in even the strongest storm. The trees around them, whipped to a frenzy, seem to be where all motion is. But ironically, in the electromagnetic activity, these steel structures are constantly in a state of stormy activity. It's an easy slippage into Bachelard's words from here: 'In a storm, too, the tree, like a sensitive antenna, sets the dramatic life of the plain into motion' (*Air and Dreams*, 216).

The spiritual colonisation of what is now known as the wheatbelt has long concerned me. As a child attending church services in the oldest inland church in Western Australia – belltower cracked by the Meckering earthquake in 1968 – I became aware of the disjunction between the ideas of place I'd received through farmers' recounting of Indigenous 'stories' and the teachings of land in church, especially in Isaiah and Ezekiel, which belonged to an entirely different landscape. However, in my head, without a political knowledge of the problems associated with doing so, I merged the two.

As a poet in adult life, with committed views on land rights and respect for the intactness of cultural coordinates outside my own, I can only deconstruct this conflation. Walwalinj is a very sacred place to the Balardong-Nyungar people. Yet they are entirely cut off from access to it due to the nature and telecommunications reserves being surrounded by farmland. Paragliding clubs have greater ease of access through private dealings with landowners than the Balardong-Nyungar community.

When I write about the mountain out of our back door, all of these factors necessarily inform my interpretations of this place. The corruption of access for traditional people seems to me to go hand-in-hand with the poison, the radiating, of the entire environment and atmosphere around the mountain's summit. The poem that follows, though written about an icon painted by the great contemporary Russian icon-painter, Alexander Deriev, is about all of these issues (see my *Icononotes* with Alexander Deriev):

Isaiah Icon

Woe to you who add house to house
and join field to field
till no space is left
and you live alone in the land.
– Isaiah 5:8

The Mount Bakewell Conservation Park.
Walwalinj Conservation Park.
Both have a certain
kind of ring to them:
as names, as possibilities.

These are not socially conscious
reviews of landscape,
but right-on-the-edge
resistances as the woe
comes to the besieged

flora and fauna, the wheatbelt
alpine orchid that exists
only up there, a strong
stone's throw away.

The farmers are joining
    paddocks; the recreationalists –
  paragliders, climbers, shooters –
    are joining erosions that supplant
in place of trees, flowering plants,

kangaroos, golden whistlers, parrots;
    building out like the trend
  for country living,
    an end in itself,
a means of resolution;

alone with the quarry,
    alone with the bird's eye view
  of a distant neighbour's little white lies,
    alone with a contemplation
of spaces in the body

closing in, filling out,
    the run-together thoughts
  blurring profit and loss,
    cancer cells setting off
biopsy alarm bells,

the ring of fate and damnation
    and claustrophobic interiors
  of conservation out of reach
    of the wanderer, the tourists…

I have already mentioned the obvious link between the telecommunications masts and trees. Taking the scriptural theme one more step, one might think of the Asherah poles, compared to, or planted beside trees, of Kings and other books of the Hebrew writings; these poles appeared on many high places and were condemned by the God of the Israelites because they pertained to a rival religion. I allude to this by way of demonstrating a long-standing Judaeo-Christian cultural concern with a particular image of verticality. In the context of the primarily Christian colonisation of this region, this seems somewhat ironic in the least. Furthermore, at Easter each year in this town, and permanently in Pingelly, another wheatbelt town, huge metal crosses are erected on local hills as a kind of symbolic telecommunication of the Passion of Christ. I have also written poems, unsurprisingly, about this phenomenon. The Cross is also known as the 'tree'.

Bachelard, talking of the image of the tree and verticality, tells us:

> from this vertical life, the most diverse imaginations, whether they be of fire, water, earth, or air, can relive their favourite themes. Some, like Schopenhauer, dream of the pine tree's subterranean life. Others, of the wind and the angry rustling of pine needles…. Thus a single object can give 'the total spectrum' of

> material imaginations. The most disparate dreams cluster around a single material image. It is even more striking to think that these disparate dreams, before a tall, upright tree give themselves over to a certain orientation. Vertical psychology imposes its primary image. (*Air and Dreams*, 205)

There are many tangents that one might take from this point of departure, whether it be looking at Roberto Juarroz's 'Vertical Poetry' as vertical textuality, or poems on trees, or human-made structures that resemble, mimic or replace trees. There are also individual poems such as Plath's 'I Am Vertical' in which the vertical is public and susceptible to scrutiny, and the horizontal becomes the private and the hidden. Or even, indeed, thinking of mimesis, how verticals constructed in particular patterns, such as telecommunications masts interlinking across the wheatbelt, mimic processions or progressions in various religious or social rituals. Eric Auerbach, in his *Mimesis*, says:

> Many realistic forms – the Dance of Death for example – have the character of processions or parades. The traditionalism of the serious, creatural realism of this period [the Middle Ages] is explained by its origin. It stems from Christian figuralism and takes almost all its intellectual and artistic motifs from the Christian tradition. (247)

I wish to reiterate that in highlighting these possible connections, I am attempting to critique the interference and overlays of a colonising process, and not to add to them. Trees, as mentioned earlier, are both vertical and horizontal at the same time. Another association, that between the tree and the gallows, is reiterated in Bruce Dawe's poem 'A Victorian Hangman Tells His Love' (see his *Sometimes Gladness*):

> This noose
> with which we're wed is something of an heirloom, the last three
> members of our holy family were wed with it, the softwood beam
> it hangs from like a lover's tree notched with their weight.

This horrifying image represents the gallows as communication device not only in its public warning, but also in the intimate conversation between those who are united in the death it brings. (The death sentence wasn't actually taken from all of Australia's law books until 1985, though the last hanging – of Ronald Ryan, in Victoria, which 'inspired' Dawe's poem – was in 1967.)

If the associations I am making seem tenuous, it is worth noting that defenders of mobile phone technology insist that radiation from these towers travels primarily horizontally, on a slight slope, with relatively little threat to those immediately below, yet of course there is vulnerability at the point where the beam reaches the ground. The whole issue of what

Figure 1. Pink Lake, 11 km outside Quairading on the road to Bruce Rock (see p. 60).

Figure 2. Standpipe near Spencer's Brook Tavern on Spencer's Brook Road, between York and Northam, Western Australia (see p. 61).

Figure 3. Disused telegraph line along Quairading to York railway, 5 km out of Quairading. Taken from road that runs alongside railway track (see p. 65).

Figure 4. Dust devil taken just outside Quairading, summer 1998 (see p. 71).

Figure 5. Walwalinj at night (see p. 78).

Figure 6. Telecommunications tower just outside Northam off the Great Eastern Highway (see p. 79).

Figure 7. John inside a gravel pit near York. Photograph by Wendy Kinsella (see p. 91).

Figure 8. Old iron ore mine in the Koolanooka Hills near Morawa (see p. 99).

Figure 9. 'I open the poem and read into the wind in this uncanny, disturbing environment' (see pp. 99, 101).

Figure 10. John in garden. Photograph by Wendy Kinsella (see p. 138).

Figure 11. Slade family tombstones in Katrine cemetery, near Toodyay (see p. 141).

Figure 12. Kangaroos approaching the western boundary of John and Tracy's place at Jam Tree Gully (see p. 148).

Figure 13. Western thornbill, Jam Tree Gully (see p. 148).

Figure 14. Jam Tree Gully bull ant (see p. 148).

constitutes verticals and horizontals is strongly intertwined with the rhythm of the signals themselves, forming an array of vertical stanzas with complex linearity across the landscape functioning as field of the page.

There is no escaping vulnerability, and many people will respond by pointing out that if it's not the radiation from phone towers that gets us, it'll be something else. Apart from being a cop-out on a personal level, this attitude suits market capitalism and the State, and is a self-fulfilling prophecy, or even an evocation of immunity from having to worry about the effect of our consumer lives on 'nature'. In discussing these issues, my partner Tracy reminds me of Geoff Page's poem, 'Out There', in which the reader can't really tell whether the speaking voice is holding a particular view or satirising it (see Page's *Seriatim*):

> I've seen what's out there way beyond
> the city lights and cars
>
> that flow like complex sentences
> too difficult to parse.
> I love the carbon compromise,
> the smell of coffee bars.

Tracy also brings to my attention Margaret Atwood's lines from *Power Politics*, in a poem discussing Western social (post-)modern estrangement of humans from 'nature': 'Do you want to be illiterate?/This is the way it is, get used to it.'

Constant questions of vulnerability and anxiety are in fact protections against corporate or State entities that would exploit us for their profit. The cyborg acceptance of vulnerability, as opposed to intactness of 'self', runs the risk of complicity with the State and its corporate extensions/supports. Donna Haraway writes:

> Immunity and invulnerability are intersecting concepts, a matter of consequence in a nuclear culture unable to accommodate the experience of death and finitude within available liberal discourse on the collective and personal individual. Life is a window of vulnerability. It seems a mistake to close it. (*Simian, Cyborgs and Women*, 224)

We hide behind these postmodern myths of vulnerability and ambiguity, and what we in fact most desire in poetry are these qualities as well, in order to allow for alternative and subtextual readings that provide for the agency and individuality of the reader. I have long defended this myself – that is, the recognition that a poem does not read the same for one person in one place in one context, as it would for others in different places and contexts. However, if one is considering a poetics of activist outcomes, in which the very prosody of the poem itself reflects upon and reflects those *politics of outcome*, then one has to consider exactly what these ambiguities and vulnerabilities might mean. Do we value the aesthetic over the desired activist outcome? Or is aesthetics in fact merely an adornment to attract as wide a range of readers to the 'cause' as is possible? Of course I am getting into very dangerous territory here, but I do believe that a desired activist outcome is not a diminution of the possibilities of a poetic text to be moving, informative or intellectually challenging, by being specific in its aims. Contrary to Haraway, the bombardment of the pleasures of the text (these are obviously not Haraway's words) and maybe by extension the creation of the cyborg-poem which incorporates both the flesh of the body and the technology that allows it to be recorded and transferred (from printing press through to text messaging through to the Internet) is a myth – there is a cost involved in all of this. Of course life does have a cost, but a consciousness of minimising this cost should be part of how and why we constitute our poems, and certainly how we convey them to our audiences.

It seems to me that the logical extension of Haraway's and others' arguments is to intersect with those of hybrid artists working with biotechnology. The living becomes expedient in the process of creating an aesthetic. An example of this practice is to be found in the SymbioticA research laboratory at the University of Western Australia where you can in fact do a Master of Science in Biological Arts (doing 'creative bioresearch'). This can entail an entire subversion of scientific ethics, even to make use of living organisms to convey artistic messages. Poetry and text are always favourite transferences, utilising vehicles such as skin or other body parts, human or otherwise. If you don't see much of a gap between this and using paper (and we have to be conscious of our usage here too), then there are at least two levels of scrutiny to consider. The first is the abuse of the rights of the animals involved, and secondly the issue of torture. The use of living animals clearly fits into this category. Probably the most famous example of what might concern us when talking about the ethics of usage and manipulation of living organisms for so-called 'artistic purposes' is the transgenic 'bunny' engineered by Eduardo Kac:

> With GFP Bunny Kac welcomes Alba, the green fluorescent rabbit, and explains that transgenic art must be created 'with great care, with acknowledgment of the complex issues at the core of the work and, above all, with a commitment to respect, nurture, and love the life thus created.' The first phase of the GFP Bunny project was completed in February 2000 with the birth of 'Alba' in Jouy-en-Josas, France. The second phase is the ongoing debate, which started with the first public announcement of Alba's birth, made by Kac in the context of the Planet Work conference, in San Francisco, on May 14, 2000. The third phase will take place when the bunny comes home to Chicago, becoming part of Kac's family and living with him from that point on.
>
> <www.ekac.org/transgenicindex.html>

If you are wondering what the connection is between this work of 'art' and poetry, I first came across Kac's work in a poetry journal, *The Notre Dame Review*, which carried his 'Biopoetry', which I will get to shortly. Before that, though, to emphasise an interconnectedness between the various issues I have been tracing here, it is worth knowing that Kac has also been eminent in the field of so-called 'telecommunications art' in the pre-Internet era. In his short poetics piece, 'Biopoetry', published in 2005, and about which I wrote to the journal's editors asking for a right of reply, only to receive a vague 'maybe', Kac includes such proposals as 'Microbot performance':

> Write and perform with a microrobot in the language of the bees, for a bee audience, in a semi-functional, semi-fictional dance.... The 'robeet' (robotic bee) would allow a poet to write a performative dance-text that has no reference in the physical world (that is, does not send bees in search of food). Instead, the new choreography (kinotation) would be (bee) its own reference. (145)

This proposed interference in the communications between bees calls to mind recent speculation (the research is not conclusive) that electromagnetic radiation might well contribute to the Colony Collapse Disorder, with loss of bee population, that is spreading through North America and some parts of Europe (see Lean and Shawcross, 'Are Mobile Phones Wiping Out Our Bees?'). This speculation has been strongly challenged since the research did not prove that mobile phones or their base stations cause the bees to desert their hives; the researchers, in a limited study, merely tested the effect upon bees of DECT (digital electronic cordless technology) radiation. However, the attempt to create a poetics out of communication/miscommunication is clearly anthropocentric and self-serving; it's unlikely to have the bees' interest (and the interest of an ecosystem that depends on them) at heart.

When I was twenty-four, I wrote a poem entitled 'The Dance Movement of Bees' (see my *Night Parrots*), which includes these lines:

> bees cluster
> in the half light
> unable to decipher
> a river broken by ferries,
> the tightening railway tracks,
> the blurring of bridges
> and skyscrapers,
> the diminished scents
> of shrinking flowers

I would like to think of this as a poem that *does* try to respect the interests of bees.

Another proposal from Eduardo Kac's Biopoetry manifesto is 'interaction' (or interference?) as follows:

> Marine mammal dialogical interaction: compose sound text by manipulating recorded parameters of pitch and frequency of dolphin communication, for a dolphin audience. Observe how a whale audience responds and vice-versa. (145)

This seems overtly abusive when it is considered there is speculation on the way in which sound can distress cetaceans, possibly even to the point of self-stranding and death, and when we know so little about how this happens. Kac's proposal constitutes a poetics of abuse or in the least, a poetics of curiosity and entertainment. Communication that seeks to oppress and control is rightly identified in human discourse as reprehensible, even in such cases where coercion is used to elicit information – for example on potential 'terrorist' activities. But there is little to protect animals outside shady self-serving animal welfare regulations, which are always open to interpretation anyway, and in the case of many universities are subject to little more than committee-based interpretation and enforcement.

A.D. Hope in his empathetic and sympathetic poem 'The Cetaceans' (see his *Selected Poetry and Prose*) appropriately pinpoints the issue. The final stanza reads:

> And so, too late perhaps, since in the past
> We killed all minds that might have matched our own,
> Now we turn desperately to this last
> Cradle of kind and hope from the unknown,
> Outcasts from life, self-doomed to be outcast,
> Crying: 'Comfort us, speak to us, say we are not alone,
> Cherish us in the wilderness we have made
> And where we wander unfriended and afraid.'

We might extrapolate from this that a desire to communicate with the cetaceans substitutes for a desire to affirm our own necessity having so often destroyed the 'other side' of a given conversation. I believe Kac and artists of his ilk are often seeking to validate their own subjectivity by constantly finding other sides of a conversation that won't necessarily challenge them. That is, a conversation on their own terms.

Artists/poets such as Kac use science only in so far as it serves their curiosity and artistic utility. Poetry easily attaches itself to pseudo-science in its anxiety to find relevance to and a readership in a changing world. This brings to mind the whole notion of risk and anxiety – that which we are prepared to put up with in order to enjoy the so-called benefits of modern living. Mobile phones fit this category. In his book *The Perception of Risk*, Paul Slovic tells us that

> Citizens of modern industrial societies are today learning a harsh and discomforting lesson – that the benefits from technology

> must be paid for not only with money but with lives ... every technological advance carries some risk of adverse side effects.
>
> Reduction of risk typically entails reduction of benefit, thus posing serious dilemmas for society.... (80)

Slovic, whose concerns are for policy and how it reacts to issues of technological risk, notes that policy-makers undertake 'risk–benefit analysis', 'an offshoot of cost–benefit analysis' (80). The model in the end always boils down to the benefits outweighing the risk. These risks, though, are not necessarily measurable in the short term and though something like mobile phones can be called the 'new asbestosis' or the 'new nicotine' time bomb, the claim of not only benefits, but of necessity, of inseparability from modern life, allows for incredible leeway regarding potential damage to people and environment. (Yet it was only a few years ago that people managed unknowingly without mobile phone technology or wireless Internet.) So we observe terms emerging such as 'precautionary principle', 'prudent avoidance principle', 'ALARA' (as low as reasonably achievable), etc.

So where does this leave us as poets and activists? I have tried to show that there is a physical link between the new technologies of communication and the writing and conveyance of poetry. These links may seem tenuous, but when you consider that you are using your phones and computers constantly both to write and to convey your poetry, and that the rhythms of typing and texting and even attaching documents will affect how and what you write, it's not as distant as it might seem on the surface. In the case of poets writing landscapes in which these towers become tree become text, it is a material fact.

If you are interested in the damage not only towers but mobile phones themselves do to humans and other living things, the research is abundant and it is not my purpose here to prove it wrong or right. It is my purpose instead to bring attention to the fact that our liberties are only 'allowed' in so far as they benefit or profit the corporate State. (The corporate State can be Marxist as much as it can be 'democratic'.)

We may not all feel that the most effective form of direct action is to take a tank to ram down mobile phone towers, as one man, a former Telstra worker, reportedly did in 2007, allegedly because of radiation poisoning received while working around the towers. However, by abandoning the technology, we would allow it to die the unnatural death it deserves. There are other prosodies that we can't see in the atmosphere that have long been with us, that we might tap into instead. I'll leave them to your imagination.

CHAPTER 8

Geodysplasia: Geographical Abnormalities & Anomalies of an Activist Poetics

I

In an ancient landscape that is largely dry and much eroded, erosion is the primary 'healing' factor – or, should I say, *covering* factor. By this I mean that when the land is damaged, say through farming or mining, the crimes of this damage are potentially covered up through further land erosion brought on by heat and winds and other climatic and environmental factors, as much as anything else. Much like the dustbowl of the mid-west of America 'covering up' the characteristics of what was lost or destroyed by the wind simply blowing the topsoil and its signatures away. The question of 'abnormality' is a fraught one. Normalcy suggests a systemisation, a taxonomy of imposition. Moreover, it might be directly associated with health, and we might more than generally assume that all things aspire to health. It is essential to consider that maybe 'health' is a relative term, and that 'health' is the dictator of aesthetics.

Aesthetics, by nature, I'd argue, always default to the preservation of a status quo in which the unbeautiful is maintained in a kind of Manichean permanence of opposition, to heighten the effect and affect of the aesthetically pleasing. Mind you, that 'aesthetically pleasing' doesn't have to go deep – unless it's botox deep, or even plastic surgery deep. Ironically, the healthy can be inverted, or diverted, into the 'unhealthy' not by its state of physical well-being, but by a recasting and inverting of the Manichean divide. A kind of Janus syndrome of exchange. So the healthy becomes the abnormal.

Take, for example, the state of ecological health in the Western Australian wheatbelt. Only a few percent of the land area of the wheatbelt remains uncleared, and of that few percent a much smaller area is undamaged by 'settler' cultures in any real sense. Most of the region is either under crop, used as pasture, has become saline from over-clearing, or is being utilised for other human purposes or activities. The remnant bushland is scattered in pockets either on private property or on small nature reserves; some is government land that is used as refuse sites or gravel pits.

Gravel pits, also known as 'borrow pits', are areas of laterite balls of rock that are used in road making and other building contexts. The extraction of the gravel is appropriately termed 'winning'. The lexical side of gravel mining is self-explanatory. Gravel pits are usually embedded in rare remnant bushland – all else cleared around it. As the years go by, the gravel extraction diminishes the remaining bushland. More eco-conscious shires have replanting programmes, but because the top-soil has been removed, there is a limit to what will grow in the excavated pit area. Some of the rarest plants in the region have been found on the edge of gravel pits. Sometimes the pits are closed down and left as nature reserves; more often than not they are mined into oblivion.

See Figure 7.

To all intents and purposes, the treatment of these areas of remnant bushland would be seen as abnormal in terms of the 'aesthetic' and 'taxonomical' intactness of the place. I juxtapose and elide these two terms specifically, as in many ways they are contradictory or at least anomalous. But a taxonomy is a form of control in definition, and 'aesthetics' imply their own, internalised dialectic. Bushland is a concept. An externalising. If you live entirely in bushland and have never had any reference point outside bushland, obviously you have no concept of it as bushland. The preservation of bushland becomes an act of differentiation from where one lives and what one needs to live on and with.

For example, in trying to set up a 'private' and 'public' nature reserve, we consider purchasing a cleared area on which to build a house: powered by solar electricity, with a composting toilet, and growing veganic veggies. As much as possible, deleting our reliance on profit making, and instead using barter and other means of exchange. Off-grid, the lexicon changes. However, we are on the 'living place' and the rest is the 'bush' – a place of conservation and preservation, whose damaged self one wishes to rehabilitate, to correct of its abnormalities. Now, from our point of view, these abnormalities are the result of the very colonising imperialist gestures that brought our families here in the mid-nineteenth century. We are part of that abnormality. And yet, the whole terms of engagement are fraught with irresolvable contradictions.

The gravel pit fits the same equation, and interestingly the kind of 'property' I have just been discussing is in gravel breakaway (ridge) territory. Now, within a concept of geodysplasia – 'dysplasia' is from the Greek *dys* (bad, disordered, abnormal) and *plassein* (to form) – gravel pits, or indeed quarries, or indeed regulating burn-off practices, also fit this construct of an abnormal growth in the land (the corruption of its cellular, or maybe molecular in the Deleuze–Guattarian abstraction, make-up – see Deleuze and Guattari, *A Thousand Plateaus*). Fire is a 'natural' part of the Australian environment but governments use fire to control undergrowth, to prevent unpredictable fires that are a risk to humans and property. There have been horrendous and massive bushfires in Australia over the years, so it's understandable, but the practice has devastated national parks and nature reserves by altering the balance of fire and regeneration that would usually take place over long periods of time.

But these are not 'growth' issues, one might respond. These are interfering with the body, maybe the corpse (if you have lost all hope) of the land, and not an issue of abnormal growth. True, but I'd extend the metaphor, as growth and death are ongoing, and not fixed in some distant point back in time. Time, especially in the sense of country in Australia, is multidirectional. We are the future's presence, as with the past. The human interaction is a growth issue, and the enforcing of abnormality part of the growth. Before recommending a cutting-out of the dysplasia, with removal of removal collapsing in its own irony of contradiction, I will point out the ultimate irony in all this. Most locals see the 'natural' bushed area as an irritating aberration, that they would clear for pasture or cropping at first opportunity. The abnormality is in the failure of their own people (whosoever they might be) to have inflicted their sense of growth on the land. So the excision is taking place, and the dysplasia of bush and gravel – usable commodity equatable with property – is being cut out. The abnormal being made normal. The fact that more clearing leads to further salinity, to further truly damaged land, is a short-term consideration. While being forced by circumstance to try tree planting and trench digging (to remove water) to stop the spread of salinity on their own land, they will clear high and low places, an action that provably results in the rise of salt. This is more than contradiction; it is paradox.

In attempting to deal with these conflicting conditions of presence and interaction with land, I have, over my life, turned to poetry with its expectations of anomaly and ambiguity. My message is clear: the health of the land is being compromised beyond hope. However, I don't believe there is such a thing as good or bad, normal or abnormal. It's always been liminal for me. So, I have a clear sense of right and wrong, but a very different sense of the terms in which these things can be discussed. I want to close all gravel pits, yet I can see even the damaged bushland as having its own beauty. The beauty of damage, and of, hopefully, repair. That's an issue of poetics that maybe ultimately separates my activism from poetry, though I have always defined myself as poet-activist. How to get around this clash of linguistic anomaly and environmental exactness?

II

Lately, I have been working on a de-dramatised version of Thomas Lovell Beddoes's verse play *Death's Jest-Book*. This renaissance-anachronistic implosion of a grim romanticism is the body and the corpse on which I wish to play these contradictions out, because its ultimate themes of betrayal seem so fitting with the abuse that colonising cultures in Australia have inflicted on the land. Their failure to hear and respect Indigenous voices, their deletions and overlayings and disrespects, have qualified the definition of bush so much (in Australia, the clichéd binary of 'city and bush' is still ever-present – the 'bush' entailing whatever is beyond the reach of the city in general) that the terms of betrayal are largely ignored or 'lost'.

In my version of Beddoes's play, characters have lost definition, and many actual phrases and the odd complete line of Beddoes have been incorporated into the text. This process of appropriation is reflective of the appropriation, carried out in Australia for over 200 years, of Indigenous cultures (there are at least 250 tribal 'countries' in Australia). The process of 'exchange' I have implemented in dealing with the Beddoes text is relevant personally, as someone who has lived for a long period of time in England, and is also familiar with Germany and Europe in general – such a big influence on Beddoes. Beddoes's literary and scientific family background, in addition to his anatomically obsessive investigations of the body, come into play – a colonisation played out in a country whose obsession with bush at least partly lies in a fear of the bush being a burial ground for people murdered in the city and taken out for secret disposal. (This is a common theme in Australian newspapers.) The fact that the land in general has been a burial zone for Aboriginal peoples murdered by the colonisers is certainly not prominent in public discourse.

In considering questions of 'abnormality', I am trying to create a synthesis between the aesthetics of creating the poem (and its 'end' result, which I always hope is deconstructive, and ultimately impermanent, and certainly changeable) and using poetry to actively stop land-damaging (and people- and animal- and plant-damaging) practices. But awareness is not enough if you don't understand the inherent contradictions in the language you use to discuss an issue or place, and its inadequacy in pinpointing what these contradictions are.

In essence, this is the poetics behind what I am doing in the *Death's Jest-Book* de-drama, extracted from an email to a friend explaining my purpose:

> it's process deconstructing bigotries in australia as much as elsewhere. to do with people and geography. it's also about (pro) indigenous landrights and tackling property issues. obviously taking on racism and othering and in other scenes tackling orientalism etc. it's also strong pro ecological rights. anti-capitalist. it is reducing then reconstituting the strains of romanticism etc. it also tackles the two cultures, esp scientific fundamentalism. and so on.
>
> however, this scene [the poem accompanying this outline] has been written specifically with the conference in mind, and the theme of geographical 'abnormality' i've set for myself. the scene is set in a gravel pit in the wheatbelt here – often the last refuges of native vegetation and animal life, in addition to being among the few places of 'public' land open to people (this has

major implications for indigenous access to traditional lands etc). i won't explain more here, but in essence you will have this piece read in a gravel pit by me and then i will provide an exploration of the 'scene'/locale with the camera's eye and provide a critical/ investigative poetics commentary for this.

III

The Beddoes text I have used is the 1829 version, as edited by Michael Bradshaw for the Carcanet imprint, Fyfield Books, in 2003.

From Death's Jest-Book Intertext (1829 text): a de-dramatisation (out of Beddoes)

Act Two, Scene IV

Unripening, unswelling plot,
snaky rebellions restless in the shell
that no longer hides them: the gravel pit
is gouged deeper by the shire, and old trees
topple on the edge of activity. A 'ruinous churchyard',
in *your* language. By moonrise I'll hide myself
in my own shadows, break the dark stagnation
of the hollows, open but lightless, the gravel glint
absorbing its own reflections, sugar melting
in a glass of poison. The spikes of dryandra bush,
regrowth when they've taken the surface,
a glistening abnormality that belongs nonetheless,
are 'native', a dysplasia like the drunk bitten
by the snake, a strange little snake
in his heart. Euphony is a latch string
pulled through surrounding trees –
the area still to be bitten, eaten out,
few years remaining – their necks
trampled and pushed aside, burnt maybes
transformed to winged stars of ash, distilled
in one thick minute. Sounding madly.
I never knew. I never knew!
Close as his blood's double-souled body;
best blood of death's holiday
they love the most: along this gravel road,
past gravel pit, small busloads of wildflower tourists:
this time of year, spring! On death's holiday
the gravel roads flesh beneath the wheels,
a brittle and mellifluous contradiction:
the driver – doubling as guide – is a wit:
each corpse, each laterite ball of gravel
screeches as we roll on over. Oxymoron.

The very garrets of grave. Whose bones
are right – it is easy to forget,
to extract, to build over. Last week
I was told of a town with a Nyungar name
that was 'avoided by the natives' – an affirmation
of settler presence. Then, today, I hear
how the shire in the early days
moved all Aborigines on – those there,
those passing on. It's all engineering, this history,
the use of gravel from gravel pits. The workers
do our bidding, the engineers
behind them. Death's commandments,
the body's citizens, stolen passports
in seas of wheat – the ears are thickening
and even the sceptics grow joyful at harvest. I do.
The smell of stalks being cut, the header greedy
for grain, spitting out chaff. Souls
have done their dash – rising, mixed with bones.
What does this mean? Unripening,
unswelling plot, snaky rebellions
in the gravel pit, its heart's heart
torn out, plain and simple, its inmost fruit,
the front-end loader's bitter
poison berry. Farmers here
still describe the field as female:
her womb, her tomb, her grave alone in paddocks,
the still-born beside her, husband cuddled up
later. Much later – murdered
in his dream of heaven, embracing
his damnation. A commonplace story?
A tale? Now, for another shape,
now for fate, now for the wheeling world.
Does he speak? Doth he speak?
Then all the minutes of my life to come
are balls of gravel in the pit, and the shadows
of surrounding trees murderesses?
Devils of abandonment? Ravishing angels?
Kissed with fire, the shire worker
digs deeper into the gravel hillside: wandoos,
marri, she-oaks, sun orchids ... all tear
and fall. 'He's having trouble with his missus',
the boss says, 'leave him to it'. Deeper
into an unknown sea. Islanded richly
with wattles and grevilleas. Snaky openings.
Passion's hieroglyphs painted on minutes
of mad thoughts. Plotter. Politician.
Grave-paved star. Raise the devil
and ask me how. His love's litany.
With the bucket of the loader
he lifts the gravel and dumps it

in the truck. He imagines
the truck's craving – suspension set
to take the weight, deserted human engine.
He thinks over banquets, funerals, weddings.
What's it been about, Death's angel? I wish
my rage to die, the gravel pit to fill
with dream hells, black intervals. The dust
already spreading. Choking, crossed with red.
Fire season is enclosing. A harvest fire
is true hate. We kick and beat the flames
as they leap fleshy walls. A ghost pinches
his love-lean body. I see his life
as in a map of rivers. He wrestles
and tumbles perilous rocks,
the breakaway. A dark haunted flood
creeps up. At night, when the machinery
is cold and still, he huddles
with snakes – garlands
for the King of the grave.

CHAPTER 9

Activist Readings of Three Australian Poems

A.D. Hope's 'The Death of the Bird' (1948)

I had to translocate myself to really think about this poem. Where we live near York, in wheatbelt Western Australia, there are many birds. Mainly birds resident the year round, but some migratory waders appear around the Avon River and farm dams, and every now and again there are irruptions of birds from elsewhere in the State. There is much that is uncanny about the place, but the uncanny dilutes with expectation and familiarity, and what has been lost to memory is reintroduced as a notion of 'the lost'; so trying to reconfigure a sense of the uncanny into the place heightens awareness.

If it had been entirely 'successful', the attempt to overlay and delete Indigenous presence through surveying, clearing, renaming, removing people, and all other tactics in the colonial modus operandi, would prevent an uncanny reconnection with the erased pre-presence. Fortunately, it has not been 'successful', and the stories of the land, the songlines and tribal associations are maintained and even reclaimed (through the teaching of language such as Nyungar) by Indigenous individuals, families, tribes, and language groups. The uncanny is part of retrieval, though in terms of the obligation of restitution I have, coming out of an early 'settler' family, I feel I have little right to a neurosis of surprise, to an experiencing of the uncanny. Of course, one can't mediate or control the uncanny – that's a contradiction – but to continue to experience it at the expense of the Indigenous 'other' whom you have directly or indirectly persecuted (in the most extreme instances, murdered) means reconciliation and restitution are still to be made. And if they were to be made, this wouldn't erase the uncanny, but surely it would change its nature.

I should say at this early point that although I am aware of Gelder and Jacob's *Uncanny Australia*, this essay is not derived from it. Freud's words 'the primordial narcissism that dominates the mental life of both child and primitive man' (*The Uncanny*, 142) lend themselves to a discussion of ways in which the 'sacred' underpins the modern (nation), which I would gather is the territory of Gelder and Jacob's book. I should also state that I reject any notion of 'primitive man', that we are not spiritually or ethically separated from primal and/or primitivist beginnings but rather only declare a separation through technologising the recording or systemising of social relations. But that's another issue.

Freud maps the presence of infant compulsiveness that accounts for later superstitious self-aggrandising, a narcissism of adulthood that personalises coincidences in relation to events. Such self-aggrandising in this occupied environment (of the WA wheatbelt, of the continent as a whole) is doubly narcissistic for the migrant, for the in-comer such as myself and all other non-Indigenous Australians. Uncanniness can exist, of course, but it carries a slippage.

In translocating myself from my 'home place' (or one of my home places), this equation isn't changed, but a lack of familiarity blended with just enough familiarity creates a sense of disconnection that allows for the uneasy, the gloomy, the ghastly (of Freud's dictionary-etymological citing) to catch one unawares before the self-critical stops it dead. The uncanny cannot last, and certainly can't be revisited once we become overly familiar with a place. This is not to say one can't have uncanny experiences there, but rather that familiarity will dilute the 'surprise', the fear (of death).

I translocated to the Koolanooka range of hills near Morawa. The name is Indigenous and means 'plentiful bush turkeys', or mallee fowl, a rare and endangered species. Heading home, we saw a mallee fowl – a ground-hugging bird that builds monumental mounds as nests to brood its young – in the middle of the road. It is said that my Uncle Jack cleared the last mallee mound from Mount Hardy near York many years ago. The translocation carries with it massive problems. York was the first inland town in Western Australia, and bears a weight of coloniality, whereas the Koolanooka range was opened up in the early part of the twentieth century to farming and, later, small-scale mining. The hills themselves were the source of the first export shipment of iron ore from Australia (to Japan) in 1966. That shipment went via the port of Geraldton, where I went to high school. As it happens, we were driving back from Geraldton the 'long way' when I got the urge to read Hope's poem aloud and to take notes on it in my journal.

The poem's uncanniness is made and undone in the massive scar of this now abandoned mine. A sign at the lookout above the pit demands that no rock object, no matter how small, be removed from the hills, since it could be an Indigenous tool, and have sacred importance. I support and agree with this – respect is essential. However, this sign obviously went up before a new mine in the neighbouring hill of the same range was opened in 2005, as another company saw that they could make mining profitable in the present economic climate. They are now extracting vast amounts of ore and sending it from this hemisphere to the next, awaiting its return as metal goods. You can't see the operational mine from where we were, though, perched over the old mine.

See Figure 8.

The old mine is stepped down to a mirror of green water – it is drought-stricken territory, though there have been recent rains. The dank mirror creates a *doppelgänger* as the mine is reflected back; it has the eerie qualities of an eye looking out of the earth up at you, a stagnant living thing in the core of the grave. It creates a double, as insurance against its own extinction, its further mining one day when ore supplies are low and that pit becomes profitable again. Given that there's also a mine that's functional, next to this defunct one (though we can't see the other one), there's another eerie doubling.

The steps here remind one of a Mayan temple (for human sacrifice) pulled upside down out of the iron hill. It's culturally an inappropriate comparison, but that's how the figurative works. The image came as a surprise. Down below, inside, was a wrecked car body, and a truck was parked near the edge, maybe sucking water abjectly out of the eye. A pair of falcons circled and caught the thermals to bend loops against the hollow. Behind the grave, and down the hill on the other side, a car park and picnic-ground of iron ore had been graded. It was there we went, and there I read 'The Death of the Bird' into a strong easterly. Our four-year-old son grew afraid that there'd be avalanches, looking at the death-piles of ore finds.

From my journal:

> *Saturday 7th July, 2007.*
>
> Out of Morawa up at the old iron ore mine (next to the 'new' iron ore mine) in the Koolanooka Hills. At picnic site #3 looking panoramically out at Bilya Rock (John Forrest – first Premier of Western Australia – established a trigonometrical surveying cairn there: the eye to occupation and ownership of all surroundings) and the stretch of salt lakes, plus remnant scrub and pastoral land. Everywhere around me, lumps and chips of iron ore. Winter dry of about 22 degrees Centigrade with blue skies lymph-like on the horizons, deepening blue overhead. And so, I open AD Hope's 'The Death of the Bird' and think about the definite articles, the specificity. I would name the species, its precise migratory route, the flora and fauna of its 'homes'. Even a red bird or a blue bird. But let's not be fooled, this is a most specific description. It is not (purely) symbolic, and not generic. It is its own centre, it is entire. It is so small and so large. Macro and micro – writing 'nature' makes you aware it always has to contain both. The sublime is in the tension between the two, never exclusive to one, despite Burke.
>
> 'The bird'. I open the poem and read into the wind in this uncanny, disturbing environment, even more disturbing because I simultaneously feel a sense of belonging, and of having been here before. I have not, though I often say of the region: 'I might

have been here as a child'. Often, that turns out to be the case. But it's my compulsion to allow for the possibility even when I know it's highly unlikely. And surely I'd remember. And then the mine would have been in full operation. The grave being dug. The funereal piles/headstones of iron being thrown up to threaten my own child decades after with avalanche.

The edge of the wheatbelt, the edge of the arid zone. The birds I have seen today – a massive wedgetail eagle in a roadside tree. We turned back and Tracy took a photo moments before it lifted into flight. Four months ago, during summer, we saw the same eagle, I am sure, and its partner, just near there perched on top of a roo carcass. We could see how massive the eagle was at ground level. This second time, it was still massive up there in the branches of a dead gimlet. Birds seen: roosts of red-tailed black cockatoos with their red-hearted tails against a black blacker than night. Dozens of them. Mulga parrots. Ring-necked parrots. Top-knot pigeons. Up here, the wind is howling up across the plains – an easterly. I will read the words of the poem and they will be ripped back in flight over my head, up towards the gaping mouth of the mine. Still, I can hear songbirds in the scrub far below – honey-eaters. In the mine, that pair of falcons still tracking. I will read stanza by stanza and note my reactions.

First Stanza: Here is between warm and cold, birds gently strong. Cars and tin roofs – sparse, few – ghost as the sun strikes, glimmers, undoes the drought-stricken lakes spread in bands.

Second Stanza: The 'speck on the map' is mined out. And yet they stay, diminished. Home is for a number of language groups. The mallee fowl attract.

Third Stanza: Memory is not 'instinct alone' here. Each generation must feed something anew to adapt to the destruction. All ghosts – the murdered, the killed – ghosts overloaded, hanging mid-air where there was a hill. Exile at home. Iron iron iron bird. Ern Malley's iron bird. De-magnetised.

Fourth Stanza: The great plains between the ranges – thin greenness, patches of wodjil. The architraves of the shifted, transplanted, transformed, 'transfigured' iron hills. The cool air off a denuded place.

Fifth Stanza: 'Custom and fear' are still here and cannot be deleted entirely. They cannot be replaced with the new fear: the gun, capitalism, 'ownership'. Love is sex after a day's work on the mines. Someone else's wife. Her place. Your place. The 'waste leagues of air' the pollution we make in our travel. So-called zeitgeist of the carbon footprint.

Sixth Stanza: The blue is friendly space though tainted. The 'inane dominion' the hollowness of colonisation. The grave. 'The bright host' is lessening – mining vehicles check us out. I think I hear an engine, movement, against the wind and against the stanza. I look back to see an arm out, sunnies looking out at me. A bright host. I turn, then glance back, and the vehicle is gone. So suddenly, and on a dangerous uncertain road. Red flags fly on all mine vehicles. Why here? Visiting the dead mine to take back to the living mine? Grave-robbers.

Seventh Stanza: It stares the bird in the face – the damage. The thin thread was broken long ago but still the magnetic qualities of the iron keep it close, attracted, bring it back. Back to where the dead now lies in its empty grave. Paradox, not symbol. Keeps it close to destruction, the new possibilities of beauty, the new beauty. The stepped insides of the hill. The 'Guiding spark of instinct' never dies and is always winking. As Indigenous artist Julie Dowling once told me: it's what you can't see but my (her) people can. The same must apply here, despite the deletion, the empty death.

Eighth Stanza: The vast design is made simple here – cleared, flattened. Hope wrote: 'There was a suggestion some years ago that the migratory birds were helped to find their way when flying at night by the lights of towns. I don't know whether it has been substantiated.' An irony to add where irony seems missing? To thwart instinct and give choice. That will defy the earth's indifference. How will they manage when the resources are spent and the lights in the towns go out? They will manage to travel and die all the same. It's a hebephrenic surface, route. Only paranoia can decode for an outsider the uncanniness of place. I feel strangely less obsessive, less compulsive here. Heimlich – home, safety. The poem's form – a shape for the words to be poured into, Pound-like, to hold up the structures when they are being pulled down to heimlich, the oldest place. Some explain it by saying they walked here in an earlier life. The explosion to break it up impacts as well as fragments, the songlines persist. 'Design' is cumulative, micro and macro.

Last Stanza: When the sun sets here, this space will be in the shadow of the death heaps. The wind howling towards the distant sea, the port. There is no malice in the earth, only in its usage and interpretation – time is profit – doubt drives people to adapt. Fear. Look in the mirror, down into the green mirror. The bird is female. The mirror is made an unfertile green in its pollution, the plains struggling to green with nurture in drought. What is being abjected here? Discharge of crops under their chemical fertilisation. The male fails here. The 'tiny burden of her death' is a contention of the female – the mining, the extraction, the conversion into property that fails. No return, no profit. This is a poem deconstructing masculinity. It is a postmodern text. Hope might have agreed if it was put to him the appropriate way. Think of 'A Blason' – the critical notes towards his apparent 'misogyny' might have been his own. He knew much more than he let on. His poems, as David Brooks suggests, need to be reread in this light. Not anti-modernist, but post-modernist. The facade of the building that's the form of the poem, the diction, syntax, all working against meaning. It is a caesarean birth, a birth with forceps.

This is followed by a poem which needs redrafting. We drove home.

See Figure 9.

In a different journal in my office over the heritage-listed York town Post Office, I made another entry on 'The Death of the Bird', this time going specifically through the poem, line by line, in an environment that, with its rumours of hauntings, some might see as uncanny, but I find as far from uncanny as is possible:

> *10th July, 2007. York.*
>
> 'The Death of the Bird'
>
> Line 1: Not every bird migrates.
>
> Line 2: 'kindles heat' – to arrive in the warmth is to cool the heat again? So only the action, the unrest drives passion and instinct. Certainty=non-stimulation. The contradiction is a generator but, again, inaccurate.
>
> Line 3: passage=colonial excursion. Warmth: at high altitudes (flying) it is cold! 'Summer station' – not the home but a vacation place. The place we rely on to keep the double of ourselves differently. Different but the same. The vacation defines what is at home, and what remains home.
>
> Line 4: Following the warmth is like serial sexual activity – pursuing climactic (and climatic) relief.... Only resolvable in (metaphysical) death.
>
> Line 5: The 'speck' is as the lights on the chart. A chart is drawn up out of experience/precedent, also collective knowledge. The enjambed 'divided' is obvious as crisis of faith and sexuality. It might make them resolvable.
>
> Line 6. John Donne – two halves. 'Summons' – no choice; controlled structure. As a machine.
>
> Line 7: But a benevolent force. As in governance? Is this his conservatism winning out? But this is, surely, self-assurance that is not really implicit. Real assurance comes through satire to make better the order (the governance)? But this is not satire. It is a struggle within a phantasm where order is to be imposed. A desperate seeking to make logical, rational. The poem is magic realism.
>
> Line 8: Failed oxymoron. Red herring. 'Summer station' is not a real home, it is a point on the map which home is not. She leaves to return to the point of her departure. Is home 'hearth' à la CJ Brennan? A paradox of un-belonging. Or is home the father looking at the mother looking at the child's eyes and feeling God might be possible? Home is birth? Does the bird nest at the 'summer station' or where it is triggered/driven from winter?
>
> Line 9: The poem begins to unwind here. It makes home of the summer station with memory being of the winter place. Thus real home is passion. It is where nurture and passion (sexual) define themselves. To 'travel' is to create comparison, to make fantasy. Thinking of Propp's *Morphology of the Folktale* – for the poem is a tale – we might ask what is the bird's motivation? Just instinct? Fate? No. Propp discusses 'motivation' –

> By motivations are meant both the reasons and the aims of personages which cause them to commit various acts. Motivations often add to a tale a completely distinctive, vivid colouring, but nevertheless motivations belong to the most inconstant and unstable elements of the tale. In addition, they represent an element less precise and definite than functions or connectives. (75)

The motivation in this poem is repetitive and not distinctive, thus (attempting to make) making the bird's death 'universal'. But is this really so? I'd argue the bird is distinctive, is not purely symbolic or generic but a taxonomy in itself. It is the bird of the cyclical tale where death is both escape and relief (domestic and sexual).

Line 10: Home and hearth (Brennan) in the retelling of what is lost – exile, immigration, colonisation, lament (death of loved ones etc). It is sustenance. We create mnemonics through the rhythm and lineation, the transposability of the text, to ensure continuity. This counterpoints the singular and fixed death of the bird. It is nodal.

Line 11: Ghosts=phantasm=outside the order=the uncanny=the certainty of unpredictability, the other-worldly extras. Whom we meet on the way in our travels (this is the anti-wanderer poem). Haunt=guilt. 'Heart's possession'=responsibility.

Line 12: Despite the patterning, it's like the Rolling Stones' 'you can't always get what you want...'. 'Home' – the strongest certainty – is never absolute. We desire security of place.

Line 13: World turned upside down. Stating the actual – Freud (via Hoffman's story) 'evil eye'/'omnipotence of thoughts'. I take this figuratively: 'We can no longer be in any doubt about where we now stand. The analysis of cases of the uncanny has led us back to the old *animistic* view of the universe, a view characterised by the idea that the world was peopled with human spirits, by the narcissistic overrating of one's own mental processes, by the omnipotence of thoughts and the technique of magic that relied on it' (Freud, *The Uncanny*, 147). 'The Death of the Bird' claims certainty and reality on one level, but also employs doubling (mirage), and investigates the narcissistic 'needs' of the bird. The bird is as much ruled by an animism as by a mechanism. This surrealist line (and the line that follows) is a mis-seeing *and* a reality.

Line 14: The palm-tree is a speck or a point crossed on the journey. Its non-ownership of its 'own' shadow is the loss of the double, a 'harbinger of death' (Freud). It is the separation of the spirit from the body: the ghost, the haunting of itself by its once-self. As Freud says, it's the case of the once familiar, now repressed: 'Now, as soon as something *happens* in our lives that seems to confirm these old, discarded beliefs, we experience a sense of the uncanny' (*The Uncanny*, 154). In these lines (13 and 14), places previously crossed/lived in/infiltrated/overlaid/repressed and remembered become the uncanny signs on the journey to death.

Line 15: Mixing spaces – spatiality overlaid and reconfigured= memory. All human aspiration towards the spiritual – the haunted – in a secular or religious context – is codified as recall and co-ordinates.

Line 16: As above.

Line 17: Death calls. Death is love because it is the rationalisation of the uncanny, as sex is a rationalisation of love. It removes the fear of the uncanny, its unsettling qualities. The poem almost can't cope with the anxiety – it shares this with Brennan's 'The Wanderer' [which I discuss below], from the collection *Poems 1913*.

Line 18: Another seemingly partial contradiction. Delicate can be urgent, but it emphasises that design is in the great and the small and all will equally fall. The poem's 'design' as opposed to chaos (of free verse?) is in crisis, though. Nature is both 'designed' and random. Tautological. This is the most neurotic of Hope's earlier poems. Despair because of release – from anticipation of death, from instinct and design? So, transfiguration – which contradicts the 'indifference' inherent in the final stanza. *Unless* earth is an aside, a stepping stone (which I don't think is the case).

Line 19: Approach of death defeats control.

Line 20: The transformative/transfigurative undoes utility, which nature is subsumed by – even the 'waste leagues of air'.

Line 21: Perspective=design. 'Inane dominions' – where there's servitude to utility (and/or the pointless dominions of death). No longer useful.

Line 22: Loss of the familial, memory thwarted, place in design negated, and 'place' – home and 'summer station' – no longer drawing her. Without them she is already dead.

Line 23: Familial/communal connection is lost. Lonely in the crowd. An enforced diagnosis of narcissism. 'Bright host' – irony of the living body then dead then elevated. The body of Christ but equally a negation. A theology and teleology of the uncanny. The living are the haunted.

Line 24: Blue – comforting. Another inversion. Living Dead. Space=freedom.

Line 25: 'Appointed Season' – by the Order of Design? God? Seasons=cycles which can't be refuted. Doesn't work in an age of climate change. Here, locally, the Nyungar people have always said there are six seasons. Now the reader induces anxiety in the text.

Line 26: The thread that guides her migration and life (her martyrdom?). Biological (DNA), spiritual, and imagined. What is instinct? Chemical/electrical, pre-configured, learnt through collective repetition. A ghosting, a phantom limb?

Line 27: *Suddenly* – the preceding part of the poem constitutes a warning. So, we have become clinical and scientific (almost), creating a nomenclature and subtextual taxonomy of the bird. The naming of 'the' bird has always induced an apprehension. Thus we God–readers–authors are deluded. All will die; 'without reason' – the poem has established 'reasons' – it doubts its own terms of reference, who and what it is. Is it the body of the female? The bird itself? It is a cyborg of text-bird-human body?

Line 28: 'winks' – acknowledgement/sarcasm/irony/awareness. The 'guiding spark' illuminates as rational transfiguration. Its mechanical purpose or process is ended, it removes itself from the cycle. Rationality leaves and death follows.

Line 29: Order gone, she desires to continue. To persist despite loss of instinct (residual memory?), or maybe it is no longer 'required'. She knows the earth, design, the system will not respond, will not empathise. It rejects what 'it'/we might interpret as narcissism, though the choice has been removed. The narcissist must be deleted. This well-ordered anarchist poem!

Line 30: Seeing in the wastes. Light=redemption – but now rationality has gone, the mechanism (or, contemporaneously, the schematics) broken, so has the sign. We *imagine* our redemption and it is absent.

Lines 31 and 32: That small wisdom, that residual instinct has left her, is not enough. Wisdom is learned, passed down, acquired. It comes from observing errors as much as successes. But the tale is a communication device, a passing on of the codes of social behaviour. It's a bleak cautionary tale here. But design takes second place to this 'human' wisdom here: it is, ironically, less complex in the poem's configuring than the complex map of hills and rivers. One is prompted to consider a possible derivation of Propp's 'double morphological meaning of a single function' (*Morphology of the Folktale*, 69). Nature's 'use' of vast design to mock the bird's small wisdom both causes the reader to question what design has meant and means in the poem overall, as well as to re-evaluate wisdom and instinct – their connectedness and relative uniqueness.

Line 33: darkness=shadow of earlier, a blotting out of the double. Sun rises and sets and vice versa as it migrates from hemisphere to hemisphere.

Line 34: The planet is now a living organism. Outside the rules of the bird's God-machine, which has deserted her. An imagined God. Home which cannot be. A loss of place.

Line 35: But – and this is another (contradictory) reversal – now the 'great earth' is without malice. Or have 'earth' and design always been separate, the former compelled by the latter as is/was the bird?

Line 36: The receiving is renewal (cyclical, seasonal, instinctual). The split personality of the poem (in theme not prosody, though there are arguments on a micro-tooled level to be made regarding this as well). Many voices and (chromatic) scales of possibility and relationship between 'beings', a co-morbidity of states of perceived and imagined being. Branching, co-existing. This poem is 'certain' in its 'executions', but uncertain in its ontology. Freud writes:

> This reminds us that this word 'heimlich' is not unambiguous, but belongs to two sets of ideas, which are not mutually contradictory, but very different from each other – the one relating to

> what is familiar and comfortable, the other to what is concealed and kept hidden. (*The Uncanny*, 132)
>
> 'The Death of the Bird' is a poem driven by revelations and occlusions.

By way of wrapping up, for there are no conclusions to be drawn, I will link the two spaces of reading by road, by driving in a car one must carefully pilot to prevent hitting birds. Birds are attracted to the wheatbelt roads by spilt grain, dead insects (wiped out in the billions by cars), warmth, and in summer, by shadow (roadside vegetation). Every drive sees numerous birds dead on the side of the road. I am careful, but every now and again am surprised, and fail. A bird's small death is writ large on *my* life, at least. Some months ago, returning from the south-west, when I was driving late at night during a lightning storm, a confused and erratic tawny frog-mouth (storms can confuse birds) flew in front of the car lights and I met it head-on. I held its warm, dead body in my hand. Its neck was broken, but otherwise it was perfectly 'intact'. Its eyes were open. In the car lights I could see myself reflected in one of its eyes. My double. Freud's harbinger of death. And indeed, I died in some way as well. It is why I write poetry. The uncanny perfection in death is 'death's harbinger'.

There are many possible interpretations of Hope's poem, but it is also about a bird in itself, about *the* bird, and about the fear of loss of home, movement, and a compulsion to persist. Reading the poem within and against the uncanny as a conscious exercise undoes, to some extent, the act. Either way, there is something fantastical about the bird, something deeply animist and empathetic. Rationality meets the fantastic head-on – a collision. As Todorov writes:

> The uncanny realizes, as we see, only one of the conditions of the fantastic: the description of certain reactions, especially of fear. It is uniquely linked to the sentiments of the characters and not to a material event defying reason. (*The Fantastic*, 47)

Fear is necessary. The reader fears the bird's fate, end or undoing as his or her own. Mostly, we are haunted by disconnection from what has gone before, a failure to lift repression of memory, and consequently our loss of identity through place.

C.J. Brennan's 'The Wanderer' (1914)

I am not a generic poet, but a landscape-specific poet who identifies and creates taxonomies. Maybe this is why I feel I get so much out of symbolist poets, who, while hyper-investing the word – especially the noun and adjective – with symbolic implication – do so starting from a generic vocabulary. The tree being of a particular species, with specific local adaptive characteristics and so on, is less important than the implication of the signifier in language and in a socio-cultural context. Even a decontextualising starts at such points, with the veil of impression holding and withholding the hermetics all such poems contain. The reader's desire for decoding, the writer's desire to weight a word beyond syntax's expectations and directions, is a desire to give meaning and relevance to 'all' that is unsaid. In my own work, I name specifically, and identify specific implications. This is not to say words don't then carry their own branching potential, their own *ostranenie*; of course they do, but they do not indicate some greater 'glory' (in the Mallarméan sense) that might be obtainable as thing in itself – they are a resonance of actuality. If this is the case, I have no desire to imitate the symbolist poets, though I do feel a compulsion to translate them.

C.J. Brennan's 'The Wanderer' is a split poem of place/anti-place and belonging/rejection. The final lines of the poem offer a resolution only in the creative self, in the dream. And here it is a liminal resolution – one in which 'dusk' is the setting, where fixity is not possible; it is 'somewhere', and even the self is paradoxically 'far in me'. The conflict between the id and the ego, however, is by no means straightforward, and the creative self and traditional self are more reconciled than conflicting. The poem could be easily mapped onto an autobiographical template, but its hermeticisms do not rely on such an exploration: the text is self-defining and self-inflecting.

I see three neuroses as pivotal to the poem. The first is fear of the dentata *and* the sexually 'unresponsive' female and female-self – either to be consumed by the maw (the fertile child-producing entry/exit) – 'a clinging home' – or to be shut out of it: 'the earth huddled her heat / within her niggard bosom'. This also manifests itself as a gender crisis over 'hearth' and 'heart', of these being tied to each other but irreconcilable. The 'wandering' urge is also licence to 'throw off the shackles', and there is an undoubted tone that will later be encountered in the '50s Beat 'on the road' libertarianism. What astonishes is how many readers have failed to link the theme of liberation from the hearth with a male discomfort at domestic responsibility, or rather the expectation that the female, muse-like, will maintain this while the creative male wanders in search of ... art and inspiration and meaning and identity. It is a poem of sublimated lust and 'lasciviousness'. It is a poem of predation and voyeurism. It looks through the windows and mocks those who can't be both inside and outside, like the speaker.

The key seems to be in section 92, which begins 'I sorrow for youth', with a lament for the loss of the maternal (the female redemptive haven) both in the wife and in the earlier mother. The lines 'but for those soft nests of time that enticed the maiden bloom / of delight and tenderness to break in delicate air' combine the motifs of ravishment and nurturing, creating the unresolvable dichotomy that impels the persona to wander, searching for what is ultimately not obtainable outside himself. The joy the persona finds in the breaking of maiden bloom is directly linked to the joy of observing the mother-to-baby gaze and desiring it for himself in permanency. This is of course unrealisable. Longing for a 'deathless heart' as haven, he says, 'where have I felt its over-brooding luminous tent / save in those eyes of delight (and ah! that they must change)'. Here there are resonances or echoes of the tent of heaven as we find in Isaiah and the symbolic pure azure skies of a Mallarmé in 'Tired of the bitter repose...'. Brennan's further neurosis over his relationship to Church and God is frequently echoed here, projected onto the female. Apart from the biblical overtones of the tent image as 'heaven as a tent' or God's dwelling, protecting and nurturing, we also have here the (equally biblical) sense of the tent as a shelter that goes with the wanderer, *his* house wherever he finds himself. It is the symbol of the nomadic with a longing for permanency, and, indeed, for God.

The second pivotal neurosis, it seems to me, is that of place and identification. As H.M. Green, in his *A History of Australian Literature*, says of the poem:

> I say literature, not Australian literature, for one thing because in spite of its physical framework there is nothing Australian about it except the birth and residence of its author; and also because it is too large to be tied down by purely Australian comparisons. The material setting of 'The Wanderer' is the long and lonely journey between Manly and Newport as it was over fifty years ago: the sea on one hand; on the other low hills and a wide sweep of bush; a cold and windy autumn evening; Brennan used to spend his weekends at Newport and he knew it all well enough. He has magnified the bush to a great forest, the low hills to a range of mountains; he could not magnify the sea. But all that is no more than a framework.... (522–23)

Green might well be right in terms of the literal physical movement from which the poem derives, but in failing to identify the neurotic nature of Brennan's generics, or 'universals', by laying them at the feet of a *revelationary symbolism* (my term), he misses the point as to why Brennan didn't use specific place references. The implication that specificity in some way limits a canvas of spiritual and creative explication, that, say, Latin or Indigenous or hermetic namings of flora and fauna reduce the universality, the spiritual ontology, is entirely challengeable. First, specificity doesn't prevent analogy, and as various L=A=N=G=U=A=G=E poets will tell us, reading and listening should be work. Also, I think generics come out of the perceived limitations of memory (which I would challenge) – that is, in the sense that a mnemonic or refrain or anaphora or any other 'aid to memory' might prove ineffective to recall specifics. Brennan is not universalising, but avoiding. Why? Anxiety about the 'legitimacy' of the culture he is part of? Perhaps. Textually, it seems more elemental. The poem is driven by a fear of not belonging, of not actually learning from what is seen on the way. To name is to lay claim. Maybe it's a poem that fears to lay claim?

Moreover, the wanderer of the poem challenges all that might challenge him, tries to establish an authority against the small ideas of the 'ordinary', and certainly the feminine. He plays with the idea of being 'king' but 'cannot tell if ever he was king / or if ever kingdoms were'.

The third neurosis is of spiritual conviction. As the wanderer trashes home, family, the mundane, God as godhead, he reconciles his loss and inability to be connected, to some extent, through nature. Soul, God, nature, become expressible as the art of the poem. The poem does not actually move in this way but repeats actions of denial compulsively. Interestingly, the poem is rhythmically calm, even when the sea 'magnifies' (as opposed to Green's claim) but remains constantly anxious. This is not a poem of stabilising or, within the greater structure of *Poems 1913*, an autobiographical 'lull' in the fury of disappointment and unhappiness. Rather, it is a systematising of anxieties, a patterning of these into a journey structure that is actually a mantra, a cyclical 'self-speaking', a poem of therapy. Wanderer motifs usually follow the pattern of some wrong done to the wanderer, or some wrong done by the wanderer leading to an unfocussed or co-ordinateless movement that entails self-discovery, catharsis, punishment, vengeance, or limbo. A non-outcome becomes an outcome. In this poem, non-belief ('if such there were') results from the tests that might apply to a believer like Job, so it becomes an anti-religious but spiritualised poem. The void, the abyss, the illumination of a visionary whiteness, or blankness, or nothingness. Darkness is held at bay through self-perception: cognitive sense is enough to keep anxiety from consuming the self.

On 6 June, in my office over the town Post Office in York, I wrote:

> I almost randomly open Todorov's *The Fantastic,* and read:
>
>> We saw that the 'themes of the self' could be interpreted as so many definitions of the relations between man and the world, of the perception–consciousness system. Here, in considering the 'themes of the other', nothing of the kind is so: if we wish to interpret the themes of the other on the same level of generality, we must say they concern, rather, the relation of man with his desire – and thereby with his unconscious. Desire and its variations, including cruelty, are so many figures representing the relations between human beings. At the same time, man's possession by what we may call his 'instinct' raises the problem of the structure of the personality, of its internal organization. If the themes of the self implied an essentially passive position, we note here, by way of distinction, a powerful *action* on the surrounding world: man no longer remains an isolated observer, he enters into a dynamic relation with other men. (139)

Interrupting the journal entry, we might note that in the fantastic of 'The Wanderer' the persona is not only considering the theme of his 'otherness', and neurotically, of the primacy of his *self-created* otherness or 'outsiderhood' but also the implied tragedy of its necessity. Choice and lack of choice are in oxymoronic struggle. The wanderer is given and claims distinction, sees his otherness and 'exile' as a powerful if fated action, and certainly embodies the romantic 'isolated observer'. Furthermore, his apparent Edenic-hunger, desiring or wish-fulfilment coupled with the constant failure to realise this (if ever, indeed, 'he' wants to), leaves him the isolated observer. Axel Clark, Brennan's biographer, talks of the gushing sentimentality of 'The Wanderer' (see Clark, *Christopher Brennan*). This I entirely reject: rather, the poem struggles with a masculine–feminine dichotomy in the passive self, and in being unable to reconcile the feminine in himself, the wanderer remains an unresolved, neurotic, and damaged figure.

As an aside, and as something to be explored further with regard to 'The Wanderer', Todorov argues that 'the fantastic can subsist only in fiction' (*The Fantastic*, 60), but his definition of poetry is narrow and genre-stricken. He notes: 'The poetic image is a combination of words, not of things, and it is pointless, even harmful, to translate this combination into sensory terms' (60). We have to step outside structuralism, but the opposite is certainly the modus operandi of many contemporary poets writing in many languages, and, what is more, I would argue that Brennan's 'The Wanderer' is a poetry of the fantastic. Even allowing for the regularities of poetic prosody, we do not need to read something as fiction (as Todorov says of Nerval's 'dreams') to induce the fantastic. I would insist that the more controlled a poem, the more we should consider where the fantastic lurks. As I note in my journal (which includes a draft of 'Un-numbered Graphology Poem: Fantastic'):

> Which brings me to think of Christopher Brennan's 'The Wanderer'. I wrote a 'wanderer' poem at Churchill College, Cambridge, a few months ago. I have been thinking of the neurotic vis-à-vis Luce Irigaray. I recently applied this to a poem by John Boyle O'Reilly, 'The Dukite Snake' [see below – NL]. The wanderer and Todorov's 'themes of the other' make a neurotic misreading necessary. As I mentioned to Barry Hill, a poet friend from Australia, Mallarmé's 1866 crisis was not religious or spiritual but domestic. There's no room in the house for 'genius'.... At home, Tracy is feeding the kids. Her 'loss' of time to the wellbeing of 'the others' is my time to ramble, grow paranoid, concentrate my neuroses. Brennan and the domestic. I don't need his biography, his 1897 marriage and what followed to see this in the poem. It screams. It is a poem only pretending to be quiet. It is louder than Lilith. Brings to mind the necessity of domestication in biographising. In autobiographising and making universal: the trick of the poet. Who are the real confessional poets? Brennan writes hungry for the domesticated, but

not the domestic. He wants to sample but not be 'trapped'. This is where the symbolic is purely romantic:

> *Quoniam cor secretum concupivi*

– or maybe memory is what enforces neuroses – loss and regaining. Like the old corner frames for photographs in albums – a broken but decisive frame. Narrative order, unsequencing the poems:

> 93 You, at whose table I have sat, some distant eve
> beside the road, and eaten and you pitied me
> to be driven an aimless way before the pitiless winds,
> how much ye have given and knew not, pitying foolishly.

'Ye' – 'he' addresses all of us. Emphatic. Retributional, God-like. Apotheosis. A replacement for the scriptures as Lilith was the first and was cast out, usurped. Back in 1991, I read Brennan's Lilith poems and wrote my own. So, 'ye' – not worth the pity. A hear-me-don't-pity-me-but-pity-me scenario? A reconfiguring of the Ego–Id tension through narrative distraction, rhetorical interlude – exchanging guilt, loss, failed self-image, with arabesques, gestures? No. That his domestic self is lampooned by his wandering self. A split. A public declaration, an oration, a soliloquy – a neurosis become paranoia. Loss, yes, but imprisonment too. We go to 'unwitting lives' and 'narrowing soul' and 'short-lived summer's mirth'. The dialogic loss and gain parry to drive out the hunger for the domestic. The initiator, the definer of the neurosis is evident in 'I know, I had it and rejected it' – the speaking for all through the ego-defining initial (Self) experience. When the romance was alive before the nitty-gritty of living and raising children and *obligation* came into play. Until reality hit.

This becomes a megalomaniacal assertion on the surface that collapses – this is yelling out of the calm of the rhythm, a manic self-revelation – a laying open, a flensing of self by kicking against accusers like a drunk does when s/he is being evicted from a pub, their home. To claim the neuroses of sexual denial, onanism, loss of the domestic through violence to the idea of others and alternatives, to declare self other, to ward off 'the other' – it is body-hating (especially female body-hating). The last two lines of 93:

> I have lived your life, that eve, as you might never live
> knowing, and pity you, if you should come to know.

That is, himself. Wander to avoid. The lines are fascinating collations of statement – axis – reply. Almost faux-saga verse structure.

The wanderer, in declaring position, is compulsive. Lament and accusation help define a position, a place that is contradictory and yet fixed in this, as he moves, giving co-ordinates in a shifting, unstable universe, through neurotic comparison, challenge, qualification etc.

Todorov writes:

> It would be tempting, at this point, to relate our second grouping, the 'themes of the other', to that other great category of mental diseases: the neuroses. A superficial rapprochement might be based on the fact that the decisive role of sexuality and its variations in the second thematic system seems, in fact, to be rediscovered in the neuroses: perversions, as has often been remarked since Freud, are the exact 'negative' of the neuroses. (*The Fantastic*, 147)

So, is there an absence of the perverse in Brennan's neurotic wanderer? No, it is also evident and the two can be co-morbid. Case studies? An early life of drug addiction. Brennan's poem neurotically indulges through announcement, and represses (textual) self-doubt through blame. There are also mild elements of psychosis in this.

Todorov continues:

> Here is how Freud approached the problem shortly after his second formulation of the structure of the psyche: 'Neurosis is the result [*Erfolg*] of a conflict between the ego and its id; whereas psychosis is the analogous result of a similar disturbance in the relations between the ego and the external world'. (*The Fantastic*, 148)

How much is denial and how much is repression in Brennan's 'The Wanderer'? How much is denial and how much repression in confronting an exchange between co-ordinates between region/ locality and an international/ist paradigm?

The love affair (the fetish!) between the ego and the id of Brennan's wanderer persona is the loved/the lost/the domestic – the GPO fixed – prevaricates between neurosis and psychosis:

> 97 What is there with you and me, that I may not forget
> but your white shapes come crowding noiselessly in my nights,
> making my sleep a flight from a thousand beckoning hands?

This supremely over-invested 'symbolist' poem is in fact a romantic poem of de-aesthetics, a neurotic and sometimes psychotic battle of self and id. It was never a symbolist poem. Its lushness of self-doubt and its obsessive repetitions (words, motifs, sounds, shape and colour/light), its anxieties of loss and belonging, leave the persona and the text itself trapped in a crisis of masculine self-possession in the struggle for ownership of the body of text (female).

> 99 The land I came thro' last was dumb with night,
> a limbo of defeated glory, a ghost.

Brennan's poetry is supremely self-analytical. The poem's release is in the acceptance of the role of the poem in the life-role of the poet moving over physical, conceptual, and emotional regions to which he/it cannot belong. It becomes a poetry of anti-nation and spiritual crisis versus the pragmatics and promulgation of place. It is a poem of self in which the unified self is replaced with a neurotic and deceiving self, struggling to avoid psychosis, struggling for a 'truth', a genuine revelation not of its own making.

99 I feel a peace fall in the heart of the winds
and a clear dusk settle, somewhere, far in me.

Postscript
Journal entry of 14 July 2007:

> Thinking of Brennan's 'The Wanderer' and reading short stories by Philip K. Dick (the greatest English-language fiction writer of the last 60 years!).... Dick's first published story, 'Beyond Lies the Wub' carries the lines:
>
> > 'I find in your Odysseus a figure common to the mythology of most self-conscious races. As I interpret it, Odysseus wanders as an individual aware of himself as such. This is the idea of separation, of separation from family and country. The process of individuation!'
> >
> > 'But Odysseus returns to his home.' Peterson looked out the port window, at the stars, endless stars, burning intently in the empty universe. 'Finally he goes home.'
> >
> > 'As must all creatures. The moment of separation is a temporary period, a brief journey of the soul. It begins, it ends. The wanderer returns to land and race...'. (*The Philip K. Dick Reader*, 5)
>
> Now, Brennan's wanderer like all wanderer-constructs is a trope, a generic and genre construct (as Todorov might say). Equally, the obvious link with Odysseus has often been noted. We might push the idea of genre further – it is science fiction. When Green talks of the possible physical location for 'The Wanderer' – that it is not mere construct of the intellect and imagination, and that it might lay claim to an 'Australianness' – I agree. But as he also quotes Hughes regarding Brennan mentioning the 'Mediterranean' in his mindset (which Green attributes to the Australian sun) and the cosmology of his work, you get a merging of temporal and mythic co-ordinates. A speculative anti-fiction or a speculative subjectivity.

John Boyle O'Reilly's 'The Dukite Snake' (c. 1873)

The Dukite Snake: A West Australian Bushman's Story

Well, mate, you've asked about a fellow
You met to-day, in a black-and-yellow
Chain-gang suit, with a peddler's pack,
Or with some such burden, strapped to his back.
Did you meet him square? No, passed you by?
Well, if you had, and had looked in his eye,
You'd have felt for your irons then and there;
For the light in his eye is a madman's glare.
Ay, mad, poor fellow! I know him well,
And if you're not sleepy just yet, I'll tell
His story, – a strange one as ever you heard
Or read; but I'll vouch for it, every word.

You just wait a minute, mate: I must see
How that damper's doing, and make some tea.
You smoke? That's good; for there's plenty of weed
In that wallaby skin. Does your horse feed
In the hobbles? Well, he's got good feed here,
And my own old bush mare won't interfere.
Done with that meat? Throw it there to the dogs,
And fling on a couple of banksia logs.

And now for the story. That man who goes
Through the bush with the pack and the convict's clothes
Has been mad for years; but he does no harm,
And our lonely settlers feel no alarm
When they see or meet him. Poor Dave Sloane
Was a settler once, and a friend of my own.
Some eight years back, in the spring of the year,
Dave came from Scotland, and settled here.
A splendid young fellow he was just then,
And one of the bravest and truest men
That I ever met: he was kind as a woman
To all who needed a friend, and no man –
Not even a convict – met with his scorn,
For David Sloane was a gentleman born.
Ay, friend, a gentleman, though it sounds queer:
There's plenty of blue blood flowing out here,
And some younger sons of your 'upper ten'
Can be met with here, first-rate bushmen.
Why, friend, *I* – Bah! curse that dog! you see
This talking so much has affected me.

Well, Sloane came here with an ax and a gun;
He bought four miles of a sandal-wood run.
This bush at that time was a lonesome place,

So lonesome the sight of a white man's face
Was a blessing, unless it came at night,
And peered in your hut, with the cunning fright
Of a runaway convict; and even they
Were welcome, for talk's sake, while they could stay.
Dave lived with me here for a while, and learned
The tricks of the bush, – how the snare was laid
In the wallaby track, how traps were made,
How 'possums and kangaroo rats were killed,
And when that was learned, I helped him to build
From mahogany slabs a good bush hut,
And showed him how sandal-wood logs were cut.
I lived up there with him days and days,
For I loved the lad for his honest ways.
I had only one fault to find: at first
Dave worked too hard; for a lad who was nursed,
As he was, in idleness, it was strange
How he cleared that sandal-wood off his range.
From the morning light till the light expired
He was always working, he never tired;
Till at length I began to think his will
Was too much settled on wealth, and still
When I looked at the lad's brown face, and eye
Clear open, my heart gave such thought the lie.
But one day – for he read my mind – he laid
His hand on my shoulder: 'Don't be afraid,'
Said he, 'that I'm seeking alone for pelf.
I work hard, friend; but 'tis not for myself.'

And he told me then, in his quiet tone,
Of a girl in Scotland, who was his own, –
His wife, – 'twas for her: 'twas all he could say,
And his clear eye brimmed as he turned away.
After that he told me the simple tale:
They had married for love, and she was to sail
For Australia when he wrote home and told
The oft-watched-for story of finding gold.

In a year he wrote, and his news was good:
He had bought some cattle and sold his wood.
He said, 'Darling, I've only a hut, – but come.'
Friend, a husband's heart is a true wife's home;
And he knew she'd come. Then he turned his hand
To make neat the house, and prepare the land
For his crops and vines; and he made that place
Put on such a smiling and homelike face,
That when she came, and he showed her round
His sandal-wood and his crops in the ground,
And spoke of the future, they cried for joy,
The husband's arm clasping his wife and boy.

Well, friend, if a little of heaven's best bliss
Ever comes from the upper world to this,
It came into that manly bushman's life,
And circled him round with the arms of his wife.
God bless that bright memory! Even to me,
A rough, lonely man, did she seem to be,
While living, an angel of God's pure love,
And now I could pray to her face above.
And David he loved her as only a man
With a heart as large as was his heart can.
I wondered how they could have lived apart,
For he was her idol, and she his heart.

Friend, there isn't much more of the tale to tell:
I was talking of angels awhile since. Well,
Now I'll change to a devil, – ay, to a devil!
You needn't start: if a spirit of evil
Ever came to this world its hate to slake
On mankind, it came as a Dukite Snake.

Like? Like the pictures you've seen of Sin,
A long red snake, – as if what was within
Was fire that gleamed through his glistening skin.
And his eyes! – if you could go down to hell
And come back to your fellows here and tell
What the fire was like, you could find no thing,
Here below on the earth, or up in the sky,
To compare it to but a Dukite's eye!

Now, mark you, these Dukites don't go alone:
There's another near when you see but one;
And beware you of killing that one you see
Without finding the other; for you may be
More than twenty miles from the spot that night,
When camped, but you're tracked by the lone Dukite,
That will follow your trail like Death or Fate,
And kill you as sure as you killed its mate!

Well, poor Dave Sloane had his young wife here
Three months, – 'twas just this time of the year.
He had teamed some sandal-wood to the Vasse,
And was homeward bound, when he saw in the grass
A long red snake: he had never been told
Of the Dukite's ways, – he jumped to the road,
And smashed its flat head with the bullock-goad!

He was proud of the red skin, so he tied
Its tail to the cart, and the snake's blood dyed
The bush on the path he followed that night.

He was early home, and the dead Dukite
Was flung at the door to be skinned next day.
At sunrise next morning he started away
To hunt up his cattle. A three hours' ride
Brought him back: he gazed on his home with pride
And joy in his heart; he jumped from his horse
And entered – to look on his young wife's corse,
And his dead child clutching its mother's clothes
As in fright; and there, as he gazed, arose
From her breast, where 'twas resting, the gleaming head
Of the terrible Dukite, as if it said,
'I've had vengeance, my foe: you took all I had.'

And so had the snake – David Sloane was mad!
I rode to his hut just by chance that night,
And there on the threshold the clear moonlight
Showed the two snakes dead. I pushed in the door
With an awful feeling of coming woe:
The dead was stretched on the moonlit floor,
The man held the hand of his wife, – his pride,
His poor life's treasure, – and crouched by her side.
O God! I sank with the weight of the blow.

I touched and called him: he heeded me not,
So I dug her grave in a quiet spot,
And lifted them both, – her boy on her breast, –
And laid them down in the shade to rest.
Then I tried to take my poor friend away,
But he cried so woefully, 'Let me stay
Till she comes again!' that I had no heart
To try to persuade him then to part
From all that was left to him here, – her grave;
So I stayed by his side that night, and, save
One heart-cutting cry, he uttered no sound, –
O God! that wail – like the wail of a hound!

'Tis six long years since I heard that cry,
But 'twill ring in my ears till the day I die.
Since that fearful night no one has heard
Poor David Sloane utter sound or word.
You have seen to-day how he always goes:
He's been given that suit of convict's clothes
By some prison officer. On his back
You noticed a load like a peddler's pack?
Well, that's what he lives for: when reason went,
Still memory lived, for the days are spent
In searching for Dukites; and year by year
That bundle of skins is growing. 'Tis clear
That the Lord out of evil some good still takes;
For he's clearing this bush of the Dukite snakes.

In his *Australian Literature* (published originally in 1940), E. Morris Miller says that John Boyle O'Reilly was the 'first poet associated with Western Australia' (204). He was in Western Australia for two years in the late 1860s. However, if we allow that there was a tradition of publishing local verse from the inception of the newspaper/journal press in the colony (from 1829 onwards), this can be seen as merely a statement about external recognition and publication; that is, recognition outside the colony. I would argue that even in terms of external recognition, there were other poets from the colony who were recognised by poets from outside Western Australia as representative of the colony prior to O'Reilly (for example, Elizabeth Deborah Brockman in the 1850s). We might lay equal claim to such a title for a poet like E.D. Brockman, who, it seems, published poetry in the eastern states (though this is hard to track down). O'Reilly's first poems about Western Australia were published (in Boston) in 1873 in book form – the same year Henry Clay published his first book of poetry in Perth.

It's worth digressing here. William Grono made the following observation in the introduction to his own anthology of Western Australian poetry, *Margins: A West Coast Selection, 1829–1988*:

> Throughout the nineteenth century many colonists took advantage of the newspapers' hospitable attitude towards poetry. Alongside poems by such popular English poets as Tennyson, Keble, Hood and Arnold, the local papers published verses by their readers, many of whom chose to remain anonymous. No doubt they were discouraged from revealing their identities by the smallness of the colony (5,000 in 1850; 30,000 in 1880) and the public attention some poems received. 'Short and Sweet', a poem in the *Perth Gazette* (16.12.1859) criticised 'The Minstrel of Ara-Glen' for prolixity. And in 1872, Henry Clay's irritating habit of writing poems advising his fellow colonists on how to behave and what to believe was lampooned in a poem, 'The Illustrious Henry Bray: A Lay from Champion Bay'. (17)

In *Australian Literature,* Miller writes of Clay, in the chapter on Western Australia:

> The first author of a separate publication of verse in Western Australia was Henry E. Clay. Born in Cheshire, England, 1844, he arrived in 1858. His book was entitled *The Year and the Book: a Diary in Verse of Scripture Themes* by H.E.C. (Perth 1876); a second was *Immortelles: the Goal of Life and other Poems* (Perth, 1880). Clay died in 1896, and his poems were collected by his niece, S.E. Booking, of Busselton (*Poems*, Lond., 1910). The themes were mainly moral and religious. (205)

I quote this in full for two reasons. First, there's the relevant intertext with the quote from William Grono's introduction regarding Clay's 'moralising', and secondly, because of a discrepancy I am about to introduce over the date and title of the publication of the first book in Western Australia, the youngest of the States of Australia. We read of Clay in an article by Alfred Carson, from a 1938 edition of *Early Days: Journal of the Royal Western Australian Historical Society*:

> The sum of his poetical works is not great. In all he published three small volumes. the first was titled 'Two and Two,' the second 'The Year and the Book' and the third 'The Goal of Life.' Each was published in a small edition and in a paper cover, and all three have long been out of print.
>
> His niece and nephew – Miss Laura Brooking and her brother – after his death, collected and collated what they conceived to be the best of his poems and short pieces, and these verses, as a token of their affection, they had subsequently published in London by Elkin Matthews, in a cloth bound volume, bearing on its cover the simple title 'Poems', by Henry E. Clay. (62)

The discrepancy in publication dates might seem trivial, but therein lies the reality of poetry publishing in Western Australia at the time. The first monograph collection was not a major occasion, not a landmark in perceptions of 'local' culture. If you look at an original copy of Henry Clay's *Two and Two*, you'll find no publication date. A bit of digging reveals its print date to be 1873, and that it is therefore open to the claim of first book. The ready availability of poetry volumes from Britain (and elsewhere) is easily confirmed through letters and ship manifests of the time, and the cultural cringe of the colonies made our more recent cultural cringe of the 1950s onwards look languid. Newspapers took pride in 'home-grown' talent (or, more accurately, immigrant talent), and even the 'native born' appellation that had started to gain credence on the east coast by the time the Swan River Colony was founded in 1829 was both less relevant and had less impact in Western Australia.

The next poetry volume published in Western Australia doesn't arrive until 1897, and is actually published in Geraldton, 500 km to the north of Perth. C.W. Andrée Hayward's collection *Along the Road to Cue and other Verses* is important in a number of ways. As Miller tells us, Hayward was editor of the *Kalgoorlie Sun* (and also the Perth *Sunday Times* as well as founder/editor of the *Murchison Times*, which no doubt explains the Geraldton printing of his poems, 'Murchison' being the name of the mid-west region of WA that includes Geraldton as its main town). He was thus closely connected with the largest influx of migrants into Western Australia in the nineteenth century – the goldrush of the 1890s. The goldfields poets – including the feisty 'Dryblower' Murphy, who, as Miller points out, was promoted by Hayward (he wrote the preface to *Jarrahland Jingles*, 1908) – were a vital part of the new Western Australian poetry 'community' around the turn of the twentieth century. Miller absurdly states that Hayward 'may be accepted as the first Western Australian poet, with some claim to merit, who published verse in book form' (205). Vive Clay and his work of some genius at least, *Two and Two*!

To return to the subject of E. Morris Miller's 'first poet' enthusiasm: though John Boyle O'Reilly only spent two years in Western Australia, his connection with the natural environment was strong. The 1860s were a decade of increasing engagement with the Australian environment as part of direct observation rather than through ideas of landscape mediated via Britain and Europe. Poets such as Charles Harpur and Henry Kendall on the east coast were engaging directly with the flora and fauna, the land forms and differences (from European versions) in very direct and often celebratory ways. The issue of othering, of instilling a mystery into this landscape, was still part and parcel of this poetry, but there was a distinct identifying of 'Australianness' as desirable, that a currency lad or lass, a 'native born' colonial, might have a connection to the roots of the country.

This concept, of course, involves a whitewashing and substituting of Indigenous presence and claims to land and space. The 'evolution' of a colonial literature that incorporates and even celebrates uniqueness and difference is a necessary part of 'ownership'. The irony is that it is a default process, rather than a conscious process, as the colonial forces are still attached to the 'mother country/ies' and seek to project 'where they've left' into 'what they've occupied', and in doing so largely fail in fully assimilating their new possession. (Those interested in the question of Australian poetry looking back to the 'mother country' or positively accepting the new might consider Judith Wright's 'double aspect' idea in her *Preoccupations in Australian Poetry*.) Breaking away from the 'old' and embracing the 'new' is the step to total occupation. While on one hand we might admire the progress of a new poetry that lays claim to its immediate surroundings, that comes at a price of exclusion. A process of hybridisation usually takes place in such circumstances whereby the people under occupation, usurped in terms of their land and 'power bases', are absorbed and assimilated into the dominating/colonising culture.

The hybridisation process usually involves the absorption of some of those cultural signifiers of the 'other' into the colonising culture in return – a kind of exchange. Thus Christianity could gain such a sure foothold in, say, the pagan Druidic areas of what is now England – through syncretism, incorporating and overlaying festivals, local religious motifs, and other symbols and mannerisms of a belief system into its own, which it substitutes and replaces (making slightly different in the process). In Australia, until recent decades, little of this hybrid exchange took place. Transliterations of Indigenous song cycles, Indigenous word usage (especially for place namings and renamings) might be found in nineteenth-century Australian poetry, but they are rare. One does come across lament for the loss of 'old ways', and a recognition of the injustices of colonisation towards Indigenous peoples, but this is more on the level of moral indignation than of respect for the equality of cultures.

John Boyle O'Reilly's relationship to Indigenous peoples in Western Australia must have been complex, given his overdetermined position as both colonised (in the Irish context) and 'coloniser' (though convict). Here's the very brief entry on his life from the *Australian Dictionary of National Biography*:

> (1844–1890), Irish revolutionist and author; enlisted in 10th hussars, really as Fenian agent; sentenced to death by court-martial, but his sentence commuted to penal servitude, 1866; escaped from Western Australia in American whaler, 1869; after cruising in Indian Ocean settled in Boston, Massachusetts; took part in O'Neill's invasion of Canada, 1870, and in organising rescue by the Catalpa of convicts in West Australia, 1876; edited the 'Pilot' and published four volumes of poems and the convict story 'Moondyne,' 1880; died at Boston, Massachusetts.
>
> < http://adbonline.anu.edu.au/biogs/A050426b.htm>

Now, there's a life! Though this analysis will remain largely outside of biography, between the poem and its creator are permeable and diaphanous membranes, as with the poem and its reader/s. O'Reilly's 'The Dukite' has an intactness about its form and content that invites challenge. I will argue that it is in fact a damaged body, that struggles to retain a psycho-physiological intactness that can't be maintained. The poem is mentioned in E. Morris Miller's *Australian Literature*, when he records a number of O'Reilly's poems 'associated with Western Australia'. Among others, Miller notes: 'Western Australia', 'The King of the Vasse', and 'The Dukite Snake'. In 1873, O'Reilly had published a volume of poems in Boston, as mentioned above, entitled *Songs from the Southern Seas and other poems*. We may consider whether the poems were composed (and remembered?) in situ, or composed later, from a distance. Miller's brief commentary on the characteristics of O'Reilly's Western Australian poems is worth noting in full:

> Some of these poems are in narrative form, describing local scenes and incidents, and others are lyrics, inspired by the beauty of the bush. 'Western Australia' records affectionate remembrance:
>
> O beauteous Southland! land of yellow air,
> That hangeth o'er thee slumbering, and doth hold
> The moveless foliage of thy valleys fair
> And wooded hills, like aureole of gold.

> O'Reilly refers to the songless birds and scentless flowers – reminiscent of Gordon – the absence of fruit trees and bees, but has hopes for fruition:
>
> O strange land, thou art virgin! thou art more
> Than fig-tree barren! Would that I could paint
> For others' eyes the glory of the shore
> Where I last saw thee; .... (*Australian Literature*, 204–5)

This is noteworthy in a number of ways. First, in terms of the non-Indigenous 'outsider's' view of the Australian bush as 'less than' the familiar European landscapes offset by an admiration of the qualities of such difference, creating a conceptual parallelism, a contrasting to build a positive out of the negatives, and to define something 'new'. Extending this, and extending Miller's observation that the poet has 'hopes of fruition', is the idea that the land might be 'civilised' into something great and worthy. To describe one of the oldest places in the world as a 'virgin' is to stretch credulity, as well as to negate the strength and power of the cultures already resident there.

Yet once again, the poet is not saying only this, but that there is more to it: that maybe European eyes are not the only eyes to see it through, or that maybe a predictably colonial viewpoint is not the only one. Since O'Reilly was in Western Australia under sentence, and not by choice, his ability still to see such positives under duress is remarkable. And furthermore, he was there as an enemy combatant, as a 'terrorist' in modern parlance – and yet the affect (and effect) of the land triumphed over the horrors of the penal system (which he noted elsewhere and prevalently). The 'moveless' and 'slumbering' notion of a benign power that needs only to be awakened is a common one in Australian poetry of the period – it has its colonial overtones, but it also applies to the power of the bush to thwart the power-base of the crown and government. That is, one can escape into the bush and start anew by tapping into this apparently latent power.

It's a complex portrait. Two side points before looking at O'Reilly's 'Dukite Snake'. First, Miller's uncritical reading of these lines is only set off by the allusion to Gordon, and possibly by extension, to Marcus Clarke's 'weird melancholy' introduction to the poems of Adam Lindsay Gordon. [Gordon (1833–70), an Englishman born in the Azores, was among the earliest 'colonial' poets to be celebrated, following his suicide in Melbourne the day after the publication of his *Bushland Ballads and Galloping Rhymes*. Perhaps not even Gordon himself had the stomach for his verse. Bizarrely, he is the only Australian poet to be honoured with a bust in Westminster Abbey. Marcus Clarke (1846–81), another Englishman, is best known for his 'nonfiction' novel *For the Term of His Natural Life*, serialised in the early 1870s and published as a book in 1874. – NL] Secondly, the ode structure of address to the land itself is 'party-lined' or 'cc'd' maybe to 'others' eyes' – raising the question of to whom the poem is addressed. The audience is presumably American (where O'Reilly lives and publishes at the time of writing, we might assume), and possibly also 'settler' Australian. Or is it to the poet from the poet? The question of for whom one writes has a huge bearing on the tone and meaning of a poem. To write out of the colonies to, say, a London audience is very different from writing to a local one (in the newspaper, for example). To write into the colonies changes the dynamic once again. Poetry is an active force. When the Hilltop Hoods speak to us through their raps, they are speaking to us here and now, and even more so in their home town in South Australia. They are also speaking to people worldwide, but the dynamic shifts and changes. Intention matters, and brings different effects.

In O'Reilly's 'Dukite Snake' the listener is offered a 'yarn', a story. The persona of the poem deploys 'mate' as a sign of familiarity, to suggest he knows us somewhat, and that we might trust him, might enter the tryst, the (mock) sacred pact of the story. Further, the 'we' has already met the 'fellow' to be discussed, and are complicit in the tale. We, the audience, are part of it. This has the bushman's song and its balladic origins tied into one, along with a ceilidh informality and feel of companionship about it. Though only in Australia for two years, as I have said, and most of that in prison (apart from his escape), O'Reilly would almost certainly have become familiar with bushmen's tales, jingles, songs, and yarns. Convicts were used as 'slave' labour in the broader community, so contact within the prison population and outside it might be assumed.

The persona finishes the first stanza with 'but I'll vouch for it, every word' – he is offering himself to us by way of trust, as part of the pact. In the second stanza the familiarity is extended to smoking and eating, to participating in the bushman's fare. It's almost nursery rhyme-ish, certainly 'yarnish'. There's no pretence and the lines and the expression are idiomatic. So much for the prelude, the stating of the system of address (the procedure); the third stanza begins the story itself. We hear of Dave Sloane from Scotland, a 'gentleman' (with an ironic aside about how much blue blood is among the settlers) who was once a regular settler but has been mad now for years. The dog is cursed at the end of the stanza and the parallelism between blue blood behaviour and the fall from grace is in gear.

The poem functions by thematic contrasts, between lines, and across lines themselves. The fourth stanza continues the tale – we read that Dave had learnt his bush ways from the narrator, that he worked harder than normal, leading the narrator to wonder if he was just wealth-hungry, and finally that he was clearing the sandalwood land not for himself but for another. Subtextually, we also learn as follows:

> This bush at that time was a lonesome place,
> So lonesome the sight of a white man's face
> Was a blessing, unless it came at night,
> And peered in your hut, with the cunning fright
> Of a runaway convict;

This suggests that non-white faces were familiar, and that runaway convicts were people to fear. Interestingly, though, the poem does not condemn the non-white or the convict. It denotes difference, but not judgement. In a paranoid-reading sense, one could suggest that in fact it does make such harsh judgements, but given that O'Reilly was a convict himself, and given that the era did not question or hesitate over overt racism, we might suggest other possibilities.

Reading on we discover that Dave has been making his 'gold' to bring his wife out from Scotland. He is working for love, and also within that common vision of the settler – for the gold that comes with hard work. As a historical comparison, George Keats, the poet John Keats's brother, had a similar vision when he went to Louisville and the banks of the Ohio River, USA, to make his fortune, as did tens of thousands of other settlers. O'Reilly's technique in telling the tale is prosaic – each stanza is a paragraph of the story – and yet he is weaving a subtext through the use of short lines and fragments of the picture. The use of couplets, occasional pairings of couplets that are so close they seem to make four rhyming lines, and even a tercet, as well as varying the stanza's rhyme pattern to indicate a turning in the storyline (discovery of dead family) builds the pace and (deceptive) certainty of the story, semi-elevating the yarn to myth. The poem is in a loose tetrameter which is often found in ballads, associated with the telling of tales, and in this case luring the reader into a false sense of comfort and familiarity.

Returning to our reading of the narrative through each stanza, in the next, the wife arrives and sees the home he's made, the settling. The happy settler family is the reconciliation of the migrant myth. The narrator then talks of the purity of the wife and he wonders how the husband could ever have borne being separated from her. The desire and longing of the (narrator) bushman subtly sexualises and threatens here, as later, in stories by Barbara Baynton and Henry Lawson, we come to expect of the Australian bush story of isolation, especially regarding women. The consciousness of this threat and sexual imbalance (availability in a place where people take what they want, clear and destroy to profit – that's the norm) is not evident to the narrator, and probably not to the poet, but it is certainly textually observable. The story is escaping its teller, as all stories do. The snake is already present in the garden of Eden! And so we read:

> Friend, there isn't much more of the tale to tell:
> I was talking of angels awhile since. Well,
> Now I'll change to a devil, – ay, to a devil!
> You needn't start: if a spirit of evil
> Ever came to this world its hate to slake
> On mankind, it came as a Dukite Snake.

Given the bulk of the poem is still to come, this is a bizarre narrative device, but a telling one. The tale is in the cause and effect, the implications of desire, aloneness and place. The use of 'Friend' is to lure you in, the reader, to make you sexually complicit in the fall from grace. The Dukite's bite is the knowledge that allows the tale to be told, and by extension becomes 'culture', and fall. The narrator takes on the role of the devil because he was always the devil. His declaration appears to protest too much. The following stanzas make the Biblical analogy implicit and active, but more than that, they humanise the snake through vengeance, a Biblical mainstay. The 'dukite' [today we would call the snake a 'dugite', presumably the English form of a Nyungar word, but the fact it was spelt and perhaps pronounced differently in colonial times merely indicates that European languages were struggling to describe – to name and therefore to colonise – a relationship to 'difference' that was so often experienced as an outrage on the imagination, for which, as we know from Kant, there simply *are* no words – NL] that has had a mate killed will exact its revenge. And that dukite, like its mate, is simultaneously presented as a loner. The contradiction of the bushman: isolation and desire, the torment of aloneness and hunger for it at the same time.

The poem twists on the paradox. The address is still to the listener (as opposed to Dave), and the warning is mixed with the threat, as all cautionary tales are. It's a poem about inevitability and fate regarding a specific spatiality. So, in the next two short, matter-of-fact stanzas we learn that Dave kills a dukite and not knowing the tale (that the dukite's mate will seek revenge) pushes hubris even further by triumphantly dragging the dead snake through the bush, inadvertently leaving a blood trail for the mate to follow, like a vein in the body of the land. This is a potent image of transgression, one in which the bush is written with the blood of the colonised, by the coloniser, the body of indigeneity and presence (perversely and yet respectfully as well) written into the knowledge–evil symbol of the snake:

> He was proud of the red skin, so he tied
> Its tail to the cart, and the snake's blood dyed
> The bush on the path he followed that night.

His punishment is extreme. He returns home, dropping the snake by the door, and the next day goes out to work, returning to find his wife and child dead, with the punishing dukite mate still there, writhing surreally out of the breast of his wife. The rape of the land has led to the rape of his wife – if not literally, at least naratorially. The codes of ownership, belonging, intactness, and self become seemingly confused, but the poem retains its jaunty, breathless pace, and the breaking of stanzas across rhymes and the composited rhymes themselves don't allow the reader to think there's any mistake or confusion. The poem is tireless in its mission:

> and there, as he gazed, arose
> From her breast, where 'twas resting, the gleaming head
> Of the terrible Dukite, as if it said,
> *'I've had vengeance, my foe: you took all I had.'*

The narrator arrives at the hut to find two dead snakes, then within the hut the dead person stretched out. The narrator exclaims, 'O God! I sank with the weight of the blow' and takes the weight of blame upon himself – the weight of the clearing, the killing (of native animals from kangaroos to snakes), for desiring what he shouldn't. A displaced mea culpa.

The polymorphous perversity of the poem increases with the 'tenderness' of the hypocritical guilty affectation of the following stanza:

> I touched and called him: he heeded me not,
> So I dug her grave in a quiet spot,
> And lifted them both, – her boy on her breast, –
> And laid them down in the shade to rest.

The touching of Dave is both a contact of mateship and a form of settler earthing, a discharging of guilt. The sexuality of the narrator (the colonising voice itself) is spent – temporarily at least – and the laying of the corpses in the ground is touching, paternal. It is anti-necrophiliac at the point when, within the range of this postmodern reading, we might expect it to be the opposite. That's why I'd argue that what is at work here is a rebellion against the colonialist nature of the period's verse in English. One might equally apply the motifs to the English treatment of Ireland, or, indeed, Scotland (primogeniture, etc.). During the night the narrator sits with distraught Dave, whose life has been sapped from him, only to hear one hound-like cry pierce the darkness. The human and animal selves, rather than exchanged, are made as one, and a 'weird' and 'mad' mutability is substituted, an animal–human reality. The genius of the poem is in the last stanza. Not only does Dave 'unbecome' in a Deleuzian sense, but he literally becomes mad and remains dumb, his last cry haunting the memory of the narrator (the rapist, the destroyer who renders destruction into art and culture).

Dave's loss of voice (possibly symbolic of destruction and theft of Indigenous space and people) with the active threat of 'madness' reveals an internal paranoia in the text that far outreaches the mechanism of Hodge and Mishra's 'externally' applied reading of subtexts in literary works [see their *Dark Side of the Dream*, which distinguishes 'oppositional' modes of post-coloniality, associated with Indigenous and other marginal groups, from 'complicit' modes associated with the white majority – NL]. As a powerful aside, at least one version of an Indigenous Wagyl myth tells of a transgressive child being struck dumb as punishment for intrusion on the sacred Serpent (known to Nyungars as the 'Wagyl' or 'Wagarl'), and reduced to making only snake sounds (see the children's book by Lorna Little and illustrator Janice Lyndon, *The Mark of the Wagarl*). Furthermore, the child has as a consequence always to look after snakes, whereas Dave in the poem 'chooses' to go on destroying them. The interweaving, substitutions, and overlayings of outcomes and interpretations increase exponentially in the last stanza.

Mad Dave is wearing a convict's uniform that the narrator presumes was given to him by some prison officer, ironically making a positive out of the negative on one hand, but emphasising imprisonment on the other – the benefactor is the imprisoner (the land 'there for the taking' will exact its revenge and dues). He is at first revealed to us only in convict clothing, as if the poem's listener is to assume he must be a convict who has gone mad; then we learn he is of noble lineage. The expectation of the fall is thus made greater: what crime could such a nobleman have committed to fall so low? This is to increase the suspense of the poem. As we eventually see in the poem, the bundle he carries on his back is an ever-increasing pile of dukite skins which indicates that there is nothing more that can be exacted from him, so even with his ongoing, constant killing of snakes, their mates can take no further revenge on him – he has already lost all.

Within the framework we have established, one is reminded of Edgar Allen Poe's 'The Raven' (1845) and the issue of release or deliverance from loss. In the tradition of the bush tale, a pragmatic end is offered in 'The Dukite' poem, an end whose irony in essence defeats the entire irony of vengeance enacted out of the wrongdoing of colonisation. Memory is a haunting.

> Still memory lived, for the days are spent
> In searching for Dukites; and year by year
> That bundle of skins is growing. 'Tis clear
> That the Lord out of evil some good still takes;
> For he's clearing this bush of the Dukite snakes.

The joke couplet or moral that provides the rough'n'ready 'bush' closure to the poem makes the tale acceptable (a sense of ironic relaxation – a pragmatically vengeful release) to the listener: lest we are all held accountable for our colonisings, in varying degrees. It allows us (and the narrator) an escape from complicity. I would argue, though, that the weight of the ending is in the inescapability of the 'memory'. Dave kills himself over each time he kills a snake, and he 'kills' the narrator as well. The poem becomes a device for self-analysis and social deconstruction, also an investigation of the unconscious. That telling 'he' of the last line can refer to either Dave, or the Lord, or both – in doing so, holding Christianity, too, up to scrutiny in the context of colonisation.

What is remarkable in reading this poem is the intentional 'limitation' of the span of its language usage. The vocabulary range is limited, and its rhymes mostly masculine. The end-stopping pronouncement that comes with these single syllable rhymes undercuts the fluidity of the narrative, driving home the narrative/self-analytical paradox of the poem. One could extend the argument from paranoid to neurotic – the poem in its protests and internalisations of male rape crisis, wrestles with the body as female (the body of the poem as female) and with questions of closure and intactness. The glib ending is almost 'confessional' of this wrestling. We might consider this in a psychoanalytic and linguistic light. As Luce Irigaray writes in *To Speak Is Never Neutral*:

> Analysis of linguistic forms decodes, behind the message the neurotic explicitly seems to want to send, another message revealing the real import of his discourse, despite the fact that the carrier of this second message is unaware of it. Beyond the impressions of a first reading or a first listening, analysis can uncover the true identity of the subject who assumes the utterance, the identity of the addressee, and the nature of the proposed object of communication. Who is speaking. To Whom? About what? (43)

So, rather than a paranoid revelation of the unintended – revelation of a subtext that the writer is unaware of but which is inevitably there because of the circumstances of writing or production – we have a neurotic reaction that, in being 'analysed' by the reader becomes a 'true identity', a revelation of who is speaking, to whom, and about what.

If we take our partial biography of O'Reilly, just in the simple fact of his being an Irish convict with anti-British associations and no doubt anti-British feelings, then we might transpose 'mother Ireland' as the ravaged land (consider Seamus Heaney's 'Act of Union' here, from the collection *North*, despite his resistance to such a reading) onto the colonisation, the claiming, the 'crowning', and the convict-isation of Western Australia (also beautiful – place-wise – to O'Reilly). Even further, given that O'Reilly was writing in English and not in Gaelic, we might think of a ravishing of the mother tongue. Having descended from a family of Irish 'settlers' who arrived in Western Australia in the 1850s, I believe this to be a valid reading or connection. Taking Irigaray again:

> It must be added that language, neuter in dictionaries and in theories, has been attributed to the mother tongue, or the language of the mother, with its sexual and social determinants, with its multiple and overdetermined economies. Especially since we are no longer in a network, or a net, of in-finite references, having no organizing centre, but rather in an interlacing of threads whose weave follows a certain pattern. (*To Speak Is Never Neutral*, 139)

What Irigaray calls 'this language of the mother' can be seen to engage with a reading of O'Reilly's poem, in the narrow lexicon of its resistance to such language through the instigation of 'masculine', 'mate' dialogue. In burying the mother and (male) child, the absolute nature of the loss is clearly identified, and also one's own culpability (as writer or reader, and sometimes both at once), but as with 'thou art more/Than fig-tree barren!' from O'Reilly's poem 'Western Australia'', another body of the female self emerges in the confrontation with the secret self.

CHAPTER 10

Working with Coral Hull on *Zoo* (A Collaboration)

Before starting the *Zoo* project with Coral Hull, I'd known her work for almost a decade – first seeing it in literary journals and at *Salt* [the publishing company, now based in the UK, that John started with Clive Newman and Chris Hamilton-Emery in the 1990s – NL], and later writing on her poetry collection *William's Mongrels* (1996) for the *Australian Review of Books*. Hull's explorations of landscape and concern for ethical issues were strong attractions for me. I use the word 'explorations' in a heightened sense here, as it is the negative aspect of human colonisation and occupation of place that she so strongly writes against. A visual poet, she makes her observations in an ethical framework that qualifies the way we see and the potential ramifications of this seeing. It makes sense that Hull is also a photographer and an artist, that there is a synthesis between words on the page and other forms of illumination. Hull's landscapes are also of the body and the unconscious – the correlation between the exploration of the land and the exploitation of the body and soul, almost a visually exploding *écriture feminine*, is intricately linked. Hers are poems in which conspiracy is actualised – they reject the politics of imposition and invasiveness. They are about the rights of the individual, of collective consciousness, of animals, of place, of different cultures. Read by some as purely emotional reactions, they are often rather logical, calculated responses to oppression, infused with a 'passion' that makes the target 'see red'. Hers is political poetry that at its best transcends the polemical. Infused with extended metaphor and a strong sense of metonymic build-up, the poems take things head-on both artistically and rhetorically.

A couple of years ago when I was visiting Sydney I met with Hull, who was going over to Taronga Park Zoo to do a Zoo Watch. She asked me to accompany her. After much deliberation – zoos to me are concentration camps (as they are to Hull), not places of preservation – I decided to accompany her and to help with her report. Both of us ended up angry and disgusted at what we found. Our collaborative book *Zoo* comes directly out of that experience. The title does not refer purely to the Taronga visit, but to human interaction with animals in general. It's the process of collaboration, and the ethical issues confronted by this work, especially as they reflect on Hull's work as a whole, that I wish to discuss.

The question of whether the process of writing poetry is more important than its ethical basis, or the subject matter that inspires and drives it, is fundamental to understanding the 'poetics' or modus operandi of Hull. Her prime concern is animal rights and the well-being of animals in general. Like myself, she is a committed vegan. She functions as an advocate for animals – a go-between for the generally insensitive and exploitative world of humans and the multiple worlds of animal life. And she is a 'doer' – a direct-action person heavily involved in Animals Rights activities, both in the tradition of the 'pamphleteer' and as a 'field operative'. Which is not to say she is unconcerned with specifically human issues. The nature of recollection and memory, of family, of the destructive nature of patriarchy, and all forms of oppression, come constantly under scrutiny in poems such as 'Praying Mantis', 'Sharpies', and 'Liverpool' – all from *William's Mongrels*.

Hull can approach such things through prose, photography, and other media with equal enthusiasm and determination. But in the end, it is the ability to create conceits, to distil the moment, and enrich through metaphor that makes all her art poetic at its core. It's not so much a case of the 'perfect words in the perfect order', but of layers of sensibilities simultaneously interacting. Romantic claims to vocation and talent are convenient and mostly class-driven. Writing out of a 'working class' ethos as conscious construct, Hull takes 'inspiration' and enhances it with necessity, even expediency. One writes because one has an obligation to do so. Even at her most 'confessional', she works solidly against the notion of the 'confessional poet'. The self and personal experience may be a focus of observation, but they are always vehicles to some 'greater' political end. Hull doesn't ditch herself, but uses herself as a mouthpiece for the unspoken and often the unspeakable.

Her process is about using the tools at hand, the most appropriate device for translating – the self. She sees the poet as having a genuine role in 'society' – making a firm contribution to the physical and spiritual well-being of the individual and the collective. For Hull, poetry is a powerful tool for highlighting, and potentially righting, multiple wrongs. This politicisation may have placed her outside usual funding environments; she has struggled for many years without the support that many ostensibly apolitical poets of equal critical stature have found more readily available.

So poetry and politics are inseparable for Hull. And this is another thing we agree on. In *Zoo*, the majority of Hull's pieces are 'direct action' in nature; most of mine work by allusion and circumlocution. I think this is what makes the collaboration function as a book. Where Hull will say something outright and then build layers of metaphorical allusion,

I'll start with the metaphorical allusion and leave the reader to distil the politics. Of course, there are exceptions on both sides, but this is a general principle. Also, Hull's pieces in *Zoo* tend to be prose poems in the main, while mine tend to be lyrical – if at times, experimental 'anti-lyrics'. We both, however, agree that genre is there to be crossed, and really the realms of prose and poetry, words and paintings, sound and sight, are indivisible.

The very act of collaboration is a political one, especially in the willingness to break the boundaries between one's own personal territory and that of another, to then take this into the public arena. Even when the poems in *Zoo* focus most on the 'self', it is in the sense that this 'self' is part of a public consciousness, that it is concerned with universal issues. The poetic voice made 'ethical'.

Over the last few years I have been involved in a number of collaborations – from text-based book-length structures based on the works of Deleuze and Guattari with the Swiss poet-sculptor Urs Jaeggi, through to email collaborations with the American poet Susan Schultz, the Australian writer and scholar Louis Armand, and the British avant-gardist Keston Sutherland. I have also collaborated on a quasi-narrative poem with John Tranter and a variety of projects – primarily based on responses to paintings – with Tracy Ryan. Most recently the volume and CD *The Kangaroo Virus* project, with sound artist and photographer Ron Sims, has been published. Different from most of these, but sharing some features with *The Kangaroo Virus* project, *Zoo* covers not only textual ground but entire worldviews. It has been as much an ethical collaboration as an artistic one – though as the book attempts to demonstrate, such divisions are false anyway. [I, too, have joined the list of John's collaborators. Among other projects, he and I have recently co-edited *Plagiarism! (From Work to* Détournement*)* and together we are currently writing *The Ballad of Moondyne Joe*. – NL]

The methodologies of a Hull poem and a Kinsella poem have some things in common. There is an accumulation of data, a series of moral and ethical references that move in and out of the text, an interplay – or meta-textual play – with the expectations of readers and audiences. The fundamental differences relate to the use of formal poetic devices, the build-up of an internal referentiality in my poems as opposed to Hull's more externalised references, and the idea that I pursue – that language can be an end in itself. Of course these are issues that arise in Hull's poetry, but there is always a specific emotional and consequently socio-political response looked for in the presentation of a Hull poem. Hull will deploy devices that lure, arrest, and eventually make the reader complicit in order to get the message across – such as conflating a child with an animal in order to stimulate a response of disgust, which is then skilfully transferred onto the object of sympathy. Persona and rhetoric are deftly manipulated to keep the reader compelled, to prevent alienation from the speaker. This is how guilt and complicity are evoked. Then a series of mnemonic triggers provides a familiar framework for the reader to position response – an incident, sight, smell, or emotion from childhood, the recollection of a story, and so on. Infused with a strong sense of visualisation – of painting – the reader is moved to respond. An atmosphere of verisimilitude is created. One of her earlier books, *The Broken Land* (1997), is a brilliant example of this cumulative process – a photo-essay, a documentary, a series of visual and emotional responses, an ethical and political manifesto, a deconstructed picaresque that confronts the spectre of death. The 'river' of this land-bound picaresque is the flow of blood from the slaughter of kangaroos, the slaughter of all animals by the machine of human exploitation.

The following poem by Hull from *Zoo* illustrates part of this cumulative process at work:

> **Fox In A Cage At The Zoo**
>
> A broken down train is reclaiming the same tunnel on a circular track. Now confronted with this living red miracle, I was intent upon making contact. Yet the red fox would have none of it. All the tantrums of my small plump legs, couldn't get that fox to stop and look. Mission had turned to obsession, down the long highway of hysteria. Her eyes shone like rich liquor, unfocused and timeless, captured like an insect in amber. Her fiery tail like a stubbed cigarette. There was no reaching into her, fingers on the wire, my chubby knees pinned to the dirt, the soft skin stapled by twigs and stones. I knew that something was very wrong here, yet I trusted the world. 'Foxes are beautiful,' I said.
>
> I suddenly felt her full of pain and worry. Nan thought the bears had frightened me, those zoochotic polar bears that swayed deep inside the concrete pits. Once my brother dropped his plastic bottle, and crawled through a barricade in his nappies to get it. The polar bears' eyes were already on the toddler that teetered on the edge. For a moment I felt the silence of nightmares that emit strained and desperate cries. This was the world of the captive bears a long way from the icefloes and Churchill. The animals were teaching me lessons. It was like a shadow crept into my heart to settle there. 'This is the world,' it said, 'welcome little girl who learns from everything.'

A more rhetorical though equally forceful piece by Hull from the *Zoo* collection is the poem in which she replaces 'cow' with 'child':

> **November 3rd, 1997, Murder Scenario**
>
> I am going to replace the subject of 'cow' with 'child,' in order to reply to your statement/question that: 'it's better to wear the skin of a dead child than to go and kill a child yourself.' First of all being a children's rights activist, this statement is offensive to me. Why? Because I think that the use of the word 'better' somehow condones the former. But aside from my gut response, I will elaborate on why I think that wearing the skin of a dead child is not a good tactic for children's rights. 1) the child is dead and has

> been murdered and you participated in the murder. (from a children's rights perspective, to wear the skin of a murdered child that you participated in murdering, whether that child is part of your family or someone else's, is unacceptable) 2) the child is dead and has been murdered by others and you did not participate. (from a children's rights perspective, to wear the skin of a murdered child is unacceptable even if someone else did the murdering) 3) the child is dead and has been found by you in a paddock and you don't know whether s/he has been murdered or not. (from a children's rights perspective, to wear the skin of a dead child is unacceptable, because you cannot be sure in a society that openly promotes the genocide of children, if that child has been tortured and murdered) 4) the child is dead and you have seen that child die, so you know that s/he hasn't been murdered. (from a children's rights perspective, to wear the skin of a dead child is unacceptable, because you cannot be sure in a society that openly promotes the torturing and genocide of children, what kind of life that child has had or if that child has been tortured. Also if you saw the child dying, I would expect that you might have tried to save his/her life) 5) the child is dead. s/he is your own child, and you know that s/he has had a good life and s/he died quite peacefully in his/her sleep. (from a children's rights perspective, to wear the skin of a dead child is still unacceptable in a society that openly promotes the torturing and genocide of children, because to do so, means that you will be advertising the skin to others who might like the look of it, and who might either support or participate, in the murdering or torturing of another child to have skin like yours. None of the above promote children's rights, therefore they are all unacceptable from a children's rights perspective.

In the *Zoo* collaboration, we as poets are responding sometimes to each other, sometimes to an ethical concern or specific event, sometimes to a general principle, but most often the poems are parallel evolving texts quite independent of each other. Many of these poems couldn't have been written without the interactions, but others would have been, and indeed were. Which made the process fascinating. The book, the end result, was the 'collaboration'. The order of texts is all important. There is no indication of who wrote what, but it is easy to pick up the individual voices and stylistic differences. These are 'universal' voices moving in, through, with, and on occasion against each other. Hull and I have different views on how much the reader or an audience can take directly – I generally feel allusion and a residual effect evoke a greater response. But generally, we exist in each other's poem, and conflate, collude, and work as a chorus line. My digressions are mainly to do with experimentation, such as in my 'echidna poems', but in the end, the concern is universal.

Which brings me back to Taronga Park. The whole business of arriving by ferry, being hauled over the circles of inferno by cable car, and being confronted by ice cream and confectionery shops set the stage. And a *stage* it was – animals in their cages and glass boxes, the absurd sideshow called 'the nocturnal house', an elephant with a genital growth, performing seals, the psychologically-wrecked apes, and so on. Coral's Zoo Watch report was damning, to put it mildly. I couldn't cope and didn't know why I was there – I felt lurid, like a voyeur of the most damaging kind. To give some meaning to it all, to make sense out of it, I proposed to Coral that we write a book about the experience.

Thus *Zoo*. Coral was at Taronga for direct-action purposes, to report on the conditions and help work towards an improvement in the animals' lot. I was there to make myself aware of what I condemn in word, what I once condemned in direct protest, but now have filed away in a series of references in my writing. *Zoo* is an attempt to redress this – to make active my protest, to frame it in such a way that there can be no equivocation about the ethical concerns at stake. But by working with Hull's more direct voice, I am enabled still to explore the poetic and linguistic concerns that have so dominated my project in recent years. I genuinely believe that it is ethically responsible to explore what it is that constitutes language, that by freeing explication, by investigating its inner workings, one can move closer to a truth. Here is one of my poems from the collaboration:

The Virtual Zoo

Each cell a pixel screened in the upper register;
what impresses the human ear but silence or the quarter-
tone of the condensed mental cycle, the psychic expedition,
those passing-the-time tricks we know prisoners
have at their fingertips? The bird's a fucking astral
traveller I reckon! Can't ya tell? Standing there
with your chips & Fanta ... those birds'-nest soup swiftlets
– Collocalia – echo locating a way out as the whole zoo
bristles like quadrophenia: cheetah, chicken, chimpanzee,
coatimundi, cockatoo, CONDOR: icon, soarer rip-roaring
thermals over kodachrome Andes from Venezuela
   to Tierra del Fuego & back again.

And a fragment from another of Hull's poems in the book:

Taronga Park Zoo in Sydney

... I will not forgive the zoo for the elephant, for his far-fetched torment. His lengthy sway in every waking moment and when he slept, thrown sideways like simulated waves inside the seal lion tank. A whiff of ocean from the harbour persisted and grew strong, as the ferries glided further and further away, whilst captive marine mammals circled in parallel swimming pools. At every cage John said, 'that's the worst.' It got worse as the day grew longer. I knew that he would not forgive them for the Andes Condor. 'It's a bird that SOARS in high altitudes!' he said, words rising up inside his

> throat. The choke of tethered emotions, and the damp finality of wings that would never meet clouds. A tremendous lonesome bird sat hunched behind layers of mesh and pipes, a wire cage six foot above its head. Taronga Park Zoo flattened the Andes. They made that condor crawl.

One aspect of Hull's work that fascinates me is her ability to make the familiar dangerous, to take the complacency implicit in our interaction and generation of popular culture and render it confronting. As we sit back and laugh at ourselves disfigured into our negative character components on television, while safely distancing ourselves at the same time, we avoid the issue. In her poem 'The Zoo Ark' Hull repeats the line 'Oh my god, they've killed Kenny' like a mantra:

> A black bear with a sugar addiction is a large problem.
> Scientists found it exciting but worrisome.
> 'Oh my god, they've killed Kenny!'
> If you could actually feel their misery.

A short sound bite from the continual flow of the television, the zoo. The uneasy laughter that comes with a splatter scene from *South Park*.

In looking for the influences of other poets on the construction of *Zoo*, one might look in all the obvious places – landscape poetry, the pastoral tradition, polemical verse, and so on – or nowhere at all. The influences on the texts may be manifold – from Judith Wright to Jacques Derrida, Les Murray to Peter Singer, Marianne Moore to J.H. Prynne, Robert Adamson to Pablo Neruda – but as a book it has arisen out of an ethical concern and its origin is in debate, investigation, and revelation. There are a myriad of other concerns subtextually explored – questions of gender identity, institutionalisation, isolation, racism, and bigotry in general. In many ways its catchcry is 'Liberty, Equality, and Fraternity' – but for all living creatures!

Hull's work strikes me as having much in common with Blake's – an element of faux-naïf of the best kind, a sense of the visionary, and a framework of moral investigation fused together in a voice that is at once hermetic and public. On the completion of *Zoo*, I look at the book as a kind of illuminated manuscript – sometimes with texts by Hull, 'engravings' by me, and at other times vice versa. Hull's is a voice in which Heaven and Hell are in constant struggle, and there is a sense of contraries. My boundaries are more blurred than this, but both of us see it as necessary to break down boundaries for some kind of truth to be envisioned. But for Hull there is a right and a wrong world, and in this she is an optimist. *Zoo* is a positive book – or, at least, a book in which the hope that wrongs can be made right is strongly expressed. For me the bird is the symbol of hope, as it is for Hull; as are all animals. For both of us, humans are the phantom limb of all animals, as they are of us.

One of the invigorating aspects of the project was the diversity of styles. As seen, Hull makes greatest use of the prose poem in block format, while I concentrated basically on the short-line lyric. But there are lyrical pieces by Hull, and a range of line-length and stanza patterns from myself. In the end, however, the subject matter dictates the form. And though writing of animals and zoos, we are not, as Hull points out, 'Zoo Poets'!

> But this is not a zoo poem.
> It sounds like a complaint.
> It sounds more like,
> I want to spoil all the fun.
> I am not a zoo poet.

CHAPTER 11

De-mapping & Reconnoitring Notions of Boundaries – Mutually Said: Blogging & Acting

['Mutually Said: Poets Vegan Anarchist Pacifist' is John and Tracy Ryan's shared blog site <http://poetsvegananarchistpacifist.blogspot.com> and the following chapter comprises some of John's revised entries from (roughly) late 2007 to late 2008. – NL]

Veganic Gardening Poetics

Time to get the garden going again. It's hard to believe that it was a high-yielding vegan wonderland only two months ago. It'd been producing for nigh-on ten months and I felt it was time to let the land lie fallow for a while. Over the next couple of weeks I will dig and hoe the weeds in, turn in dry compost, water it to let the weeds emerge then turn them back into the soil after a week or two as green compost. I'll leave it sit for another week, water, turn it again, then plant with 'original' organic/veganic seed.

It is an entirely veganic/organic garden, with no animal manures or products, no pesticides or herbicides or artificial fertilisers. The soil is quite acidic, so I do use plenty of wood ash. The wood ash itself comes from our bush stove which we run during winter, burning only storm-felled wood. We do not 'harvest' wood that has obviously settled as habitat for insects and animals.

Last summer's tomato crop (planted late) was still yielding in June! Even the Chinese cabbages were going into late autumn. The frost finishes things off when winter really sets in, but the dramatic seasonal shift brought on by climate change means long late summers and autumns. We also had a sizeable harvest of butternut pumpkins, masses of snow peas through winter, a winter and autumn continual harvest of sugar-snap peas and broad beans, excellent winter broccoli, silver beet throughout the year, even in the middle of winter, late carrots and beetroot, autumn chillies and capsicum, plus various herbs (exquisite oregano and basil).

See Figure 10.

I went out to survey the garden beds today – it's not so hot, reaching only the low 30s. A couple of weeks back it was in the mid-to-high 40s! That's around and above 120 Fahrenheit. Sears gardens. Anyway, I checked out the ground ... dry and dusty with a jungle of wild oats that came up late and were left, obscuring all. Amazing after months of no water that some things that weren't harvested, because they didn't finish in time, or weren't quite up to scratch, still struggle on. Especially cabbages. Lots of dried and split broad-bean pods, spilling beans on to the ground. Will collect those. A bean so sun-affected that it caught my eye, had gone from its regular brown opacity to a transparent redness, like a fierce eye glowing at night.

I took a couple of photos of part of the garden. When it's thriving, it attracts a large numbers of birds – all are welcome. They work out insect issues (insects are also welcome), and also naturally fertilise the ground. If it falls there without human intervention, I'm okay with it!

When I am working out there on warm evenings, willie wagtails hang around, and often a pair of red-capped robins. The twenty-eight parrots [not a number but an onomatopoeic name, approximating the sound the birds make: *twenty-eight, twenty-eight* – NL] have eaten most of the seed from the vegetables I left to go to seed. I harvest enough and they get the rest.

The garden area is quite large and spreads along the flat. It will take a lot of work to prepare but it yields enough to feed two or three families daily, once it's up and running. If I had my way, I would spend my life growing organic vegetables for a community. Maybe down the track! Oh, that garden is the source of many poems. I have a garlic poem from six months ago I haven't published yet. I include it below. A veganic garden is a poetics.

Planting Garlic During a Dry Winter to Ward Off

Break the corms –
separate cling-wrapped
inner growth, free succulent
cloves, organs that open to dry
winter air. Make happen.

Cloves, curved as half-moons
out from the turntable of roots,
rotating out of the stalk core,
unsteady walking stick,
nervous maypole.

Give of tension, therapy,
cure-all for colds and plague,
torn from the dry cradle,
broken-skinned, into
the earth's crumbling

encapsulation. Purple-
veined, as if all paper written on
still pulses with life. Ink blood
we read against, white intrusion
of page, of gleam.

An action less
than snapped apart, no more
set adrift, upright from parameter
to tip, mimic-thrust from boldness
to discrete finger up,

pointing. Thrust down, cold
to curve away from straight and narrow,
rupture surface, displace
fate. Attract rain.
Make warmth.

The sun, low, insists
its sweep across the winter dry,
stenographer's growth,
politicising nurture,
meliorating.

Poets and 'Their' Words?

Another fine day in the central wheatbelt. Winter. Further north, farmers are direct-drilling into dust. Hoping the rains will come. Wrote a piece on drought today. Thinking about land and ownership, I came across an incredible poem by Norman Cameron, entitled 'The Invader', from his *Collected Poems 1905–1953*. The last stanza goes:

Invader-outcast of all lands,
He lives condemned to gorge and crave,
To foul his feast with his own hands:
At once the oppressor and the slave.

Born in 1905, Cameron was a propagandist for the British military during the Second World War. Prior to and after the war he worked in advertising and this poem is (ironically to me) so effective because of the neat lines that accumulate, compile, like advertising slogans. What I like about it is its universality: could equally apply to the British and others invading

Terra Australis and dispossessing the Indigenous peoples, or to an invasion in, say, wartime Europe. There's also the resistance of the invaded – that no matter how hard the invader tries, they can't access the identity, the spirit of those robbed. This brings to mind issues of hybridity and, I admit a little obscurely, copyright.

As an anarchist, I see copyright as a defensive reaction of the State. To own and control words, no matter whose those words are, is to control reception and usage. As a writer, I make my living in part, and sometimes entirely, from selling my words. This is complicit with the State and capitalism in a variety of ways, though one always argues that it's better than other ways of feeding the State-capitalist system. I guess, depending on what your words are intending to do, and what they are actually doing (impossible to say, really). In terms of 'umbrella' anarchism, you can turn this process against the capitalist State – writing against it, and at the same time managing to feed yourself and your dependants. It becomes a subversion of the means of production. As I said to a correspondent today, referring to a portal where people might access poetry free unless they wish to download it for anthologising, academic and similar purposes:

> i am not one for copyright as an absolute (or as such), but i am one for poets making a living! so, on that level, i think it's very positive that poets get some reward for their efforts. textual piracy doesn't bother me, but i do like the idea of poets being fed!! ... i particularly like the fact that people can access free – then, if they wish to profit from the work, the poet gets something back – that seems ethical to me.

Intellectual property seems like another way of staking claims when all the land has been stolen, 'used up', staked and demarcated. But in the end, what's being discussed in the case referred to above is the issue of 'work'. For me, work and not property is the vital variable. Work doesn't mean one should have more rights than any other to something (an object, a space, an idea), but that appreciation and respect should be given for that work. 'Work' is not merely labour in the obvious sense; it is not only value-adding, sustaining at best and fetishising at worst, but a sense of belonging and participation. This can be cultural work. This can be spiritual work. It can be preparing the meal, or growing food, or seeking to 'protect' a piece of vulnerable bushland against the profiteers of the capitalist State, or of the State in general (communist states are just hyper-controlled and cartelled versions of capitalist states – the corporation is the Party, rather than the oligarchy of business interests).

Which brings me to think of landowners in the nineteenth century around here. One prominent family, the Slades, owned a property known as Glenavon out near Toodyay on the Avon River. We often drive out that way, and visit a small church off an elbow of the river where the Slades are buried. I have written a number of poems about the place. The Slades interest me, as they were the parents of a very great Western Australian poet, Elizabeth Deborah Brockman. There is an essay on Brockman's poetry and life in my book that's launched next week – *Contrary Rhetoric: Lectures on Landscape and Language* (2008). I have been fascinated by Brockman's slim oeuvre for many years and would go so far as to say that she is one of the greatest poets in English of the nineteenth century. She is basically unknown, even in Western Australia.

See Figure 11.

For the last few years I have been preparing a book of Deborah Brockman's work (I am told by a family member I met at a literary function in Perth that she preferred to be called just 'Deborah', and not Elizabeth Deborah), with help from my mother Wendy Kinsella, and more help to come from Tracy. It's a major task. I feel there are many undiscovered poems, and trying to track them is not easy. Almost everything we have of hers is from her book of poems published posthumously by her uncle in Scotland, *Poems* (1915). Those poems were collated from Western Australian newspapers and journals, especially the *Church of England Magazine*. What makes it more complex is that Brockman often published under the name 'E', and quite a number of poets did just that throughout the colonies. I believe some of the pseudonymous poems carried by other colonial newspapers in Australia may have been her work. I know she corresponded at least once with Henry Kendall, and swapped poems. Anyway, Tracy is going to do a few days' research soon in various archives and hopefully we can unearth a few more pieces to include in the book.

Wheatbelt Gothic – An Aside

At the Wheatbelt Festival held at Northam recently, I was asked by the brilliant and much censored artist Tim Burns, who works out of York (and with whom I hope to work in the future), what are the origins of my Wheatbelt Gothic? He wondered about American writers ranging from Poe through Tennessee Williams to Carson McCullers. I wondered about recent American films like *The Three Burials of Melquiades Estrada* (2005) or *No Country for Old Men* (2007), which came many years after I first used the expression 'Wheatbelt Gothic', but are now certainly at the back of my mind when talking about the potential for (American) film 'influence' on an Australian Wheatbelt Gothic, as much as say Bunuel and Dali's *Un Chien Andalou* (1929) imagined on a wheat and sheep farm near the Meckering faultline....

While Poe is a definitive influence on my idea of a Wheatbelt Gothic, and certainly Poe via Baudelaire, both Tracy and I responded (Tracy had just read aloud to the audience a Dorothy Hewett poem, at my request) that it surely sources in the work of Dorothy Hewett. This was no doubt originally subconscious in me, but as I have spent my life (from the time

I was a small boy – my Mum was taught by Dorothy at university) woven through Dorothy's work, I think this is true – certainly from the problematics of 'Legend of The Green Country' through to 'Country Idyll' (see her *Collected Poems*), which was the poem Tracy read out at the Festival. But when I originally wrote 'Wheatbelt Gothic or Discovering a Wyeth' back in 1992 (it appeared in my book *Full Fathom Five* the following year), it came out of being on the farm, walking in paddocks distant from the house, during the evening, following the fence line back as darkness fell, and literally hallucinating galleries of 'Western Art' – from Wyeth through to Hopper through to Gothic and even Gothic revival architecture. In the Wyeth original, 'Christina' has a longing for the house she is struggling to reach, but inevitably will reach, in its barren field. In writing my poem, I was more concerned about the denuded landscape and the intrusion of the coloniser – house in the painting, 'property' and the materials of farming in my poem (fences, sheep run, etc.) – than about the willpower of the individuated 'figure in the landscape' (the woman in the Wyeth, and also in my poem). It's a poem lamenting and critiquing dispossession rather than the passions and drives of 'possession'. Thus *ubi sunt motif....*

This will sound odd, but back then I was still suffering heavy bouts of flashbacks, and had learnt to control the cascading colours by using an aesthetic framework – poems, paintings, films, whatever. The tension between a haunting (as style, as reality) and an almost perverse spiritual affirmation was unresolvable. The people in my life formed part of the images, as did the reality of wheatbelt farming methods, land damage (and attempts to restore land), the conflict between condemning the colonised and being part of the colonisation, and an overwhelming consciousness that the land was marked in ways I couldn't see or determine but intuitively or maybe intellectually knew.

Wheatbelt Gothic is style built out of damage and theft, a struggle to keep 'good order' when the 'natural' rebels and resists one's presence with all possible means and contradictions. Wheatbelt Gothic is a restoration of agency within the lies of text. It is an attempt to awaken from a nightmare of Western colonisation and destructiveness that is part of who the/an observer might be. It can only be written out of a resistance to colonial heritage. A poem is a piece of propaganda – it is not a reality. It is made lifeless through its own style. Wheatbelt Gothic is a recognition of this, an attempt to restore vibrancy through recognising and participating in decay.

The wheatbelt is a large region and yet it is specific and varied in so many ways. The moment one uses a catch-all expression, it diminishes the micro. Wheatbelt Gothic is an ironising of its own generalities and thus the poems concentrate so much on specifics. Types of, precise points in time and space. The generative force behind it is, however, phantasmagorical. Sense is in superstition and fear, and not in the confidence of feeling good when the crops are glowing green. Real people are nowhere to be seen but they make it what it is, and receive its bounty and desiccations. The sheep are drenched and intestines cleansed.

So much poetry does this and convinces us it's best for our health. It's not so much that I can't celebrate, it's just that I want to know what the implications of such a celebration are. Yes, Tim, as Proudhon noted, 'Property is theft!' And so much poetry, art, and music are theft as well. The most 'original' work is often the most property-like. We can only be custodians, and it is incumbent on all to recognise larger, more concentrated, and more defined custodianships. Wheatbelt Gothic is a style that allows for an observation of these considerations – it has no materiality, no claim.

Wheatbelt Gothic or Discovering a Wyeth

Outflanked by the sheep run, wild oats
dry and riotous, barbed wire bleeding rust
over fence posts, even quartz chunks
flaking with a lime canker, the theme
chooses itself: ubi sunt motif, but the verse
becomes as deceptive as an idle plough,
or a mat of hay spread over the ooze
of a dead sheep that is the floor
of the soak (blood-black beneath the skin,
bones honeycombed), crystallised with salt.

And yesteryear occluded by the viscous waters
of the stone-walled well which (on higher
ground) marks the dryness of the soak as either
delusion or lie. Only green shoots hidden
in the dead sheaths of reeds on the soak's rim
hint that water supports this travesty.

And the moon absorbs the sun, its fabric
subtlety – the undressing of a summer landscape
too blond for its own good, too much an extract:
the mid-West Gothic of a lone tree stump
that appears to beckon in its loneliness – open space
as collusive as a vaulted cathedral in Europe,
and the well as much a receptacle of guilt
as the cathedral's font. And consider the potential,
no, consider the necessity, of a flaxen-haired girl
merging in this field of vision and then erupting
from a point above the waterline, the tree stump.

A Few Years Ago

A few years ago, after the publication of my book *The New Arcadia* (2005), I wrote an article entitled 'By Any Other Name' (it originally appeared in the Australian Poets' Union journal and somewhere else I don't recall as well). This article considered the poetic dynamics of local namings for plants and animals. I am fascinated by how one species of, say, a tree can be given alternative names within a region; how throughout one's life a 'misnaming' from childhood can stick and become a reality.

In my book *Divine Comedy: Journeys Through a Regional Geography* (2008), this is particularly true when it comes to the name for *Tiliqua rugosa*. Most commonly named the 'bobtail', this skink is found throughout Australia and is known in the suburbs as much as the country. Around where we live in the wheatbelt, it goes by a variety of names: yourin (the Nyungar name), bobtail, bobtail goanna, bobtail lizard, blue-tongue lizard, blue tongue goanna, and shingleback. I have heard all these names used many times. I read that they are also known as 'sleepy lizard' in the region, but I have not heard that name used by locals. Our family usually call them 'bobbies', 'bobtails', 'blue tongues', or 'blue tongue goannas'. Of course, *Tiliqua rugosa* (and I only use the Latin name as a 'constant' for convenience, not as a recognition that it is a standard that cuts across, say, the constant of the Nyungar name, 'yourin') is not a goanna but a skink. So already in a number of the namings there is a taxonomical error. From a poetic point of view, in terms of what they say about locality and issues of belonging, I find such misnomers generative and deconstructive of hierarchies of classification. The reptile scientifically called *Tiliqua rugosa* 'is' what it is to those who live with it, for an endless variety of reasons. The (mis)naming grants agency (as it should) and distinguishes an individual relationship. So when I come to write a poem, the process of naming becomes all-important. It changes the geography of the poem's language itself.

There is also another species of the genus *Tiliqua* – the Western blue tongue (*Tiliqua occipitalis*), a different lizard entirely (it is somewhat similar to look at, though without the thick scales and with a more tapered tail). Both the bobtail and the Western blue tongue have a distinct blue tongue they project with mouth agape when threatened. When I write a poem, I take this knowledge to the process of naming. Also, a lot of my poetry is 'in-situ' poetry, with notes for poems taken within the space they describe, and a specific observation or occurrence behind their writing. I am there, in the poem, with the action. 'Look, there is a blue-tongued goanna' may be technically wrong, and I know it, but that's what was said at the time, so it sticks, and the poem evolves out of the process of engagement and naming.

As an activist, I see my poems as forms of affirmation and resistance at once. I always take the side of animals and plants. I am opposed to the violence that some motorists, for example, express towards blue tongues, snakes, rabbits, foxes and other animals, by intentionally running them over – I regularly see drivers swerve to the other side of the road in order to hit creatures.

On the subject of reptiles, we saw a Western granite worm lizard a few days ago and I wrote the following poem (a 64-syllable plus 4-syllable poem!):

Western Granite Worm Lizard

Springs out of the fossorial
so tightly wound about the loop
of scale and flesh, its shovel nose
– foot of compass – lifts, distracts;
squeaking high above slenderness,
vocal cord that knots in dashes,
mesmerised in its magic box,
a binary of ones and noughts:
    eyes are lidless.

Pre Post-Net Declaration

Well, it's a long time between blog entries for me. I've got to get a number of poems up on the blog to link with a Niall Lucy essay ('Running On') that references them, and that is just appearing in the journal *Derrida Today*, published by Edinburgh University Press, so I might as well say a few words at the same time. I have been doing a lot of rethinking lately – not, as some neocon critics would like to see, to reform myself into a trudging maker of poem-artefacts that echo with creative-writing-encoded purity, but rather as an extension of my views on environment and textual responsibility.

Some time early next year, I will be going off-line pretty well. I will have to maintain some computer access through my university research fellowship, and to communicate with students, but otherwise I am reverting to snail mail and a much diminished use of computer technology. I have been writing poetry in pen and on a manual typewriter since I was a kid, and once wrote entirely that way. I started using computers at a very young age and, as I've discussed elsewhere (see 'Geography Is History?'), used the first computer installed in a school in Western Australia (during the school holidays while it was being set up – in fact, I helped set it up), and started using computers for word processing back in the eighties. Anyway, it's time to step off the techno bandwagon and revert.

A post-Luddite reaction? No, I am not advocating the smashing of computers (or anything), but ignoring them – the new weaving machines will simply vanish through neglect. I recently discovered, through Tracy's genealogical research, that my great-great-great-great-grandfather, and his father before him, and so on, back on my mother's side, seem to have been frame-makers for weavers. The French heritage in the family comes via the frame-makers. Or maybe it was Flemish?

So, off-line (largely), I guess I will tap Tracy for news of Web goings on as she will be remaining on-line. Next year, we hope to convert entirely to solar electricity for all household functions, and will likely have our own water supply. I will have a second manual typewriter and hope to start making my own paper of a reasonable enough quality to type on (and to last). I made paper with a mate back in the late eighties, but haven't attempted to do so since then. As it's law to keep the firebreaks done, I intend to mow them fairly wide and use the extensive cuttings from that as fibre for the paper making. This shift in my relationship with the broader world will make a difference in my 'literary life' as well.

Some friends see it as a new form of eremitism, especially when combined with my abandonment of air travel (I will only now travel by car, bus, train or freighter – whichever is feasible and the less environmentally exacting), and reduction of travel in general. Tracy and I have even made steps towards reducing car usage in moving between country home and university (a long drive), but that's another story. Maybe Tracy will talk about that on here next year. We're adapting a different approach to movement, let's say. There's a long way to go, and it's fraught with contradictions, but steps have to be made. It's all very well writing about an ecology of place and of text (for that matter), but for me it's essential to carry through to the pragmatics of day-to-day living.

A Poetics of Gradients

Tracy did a wonderful translation of Jacques Brel's '*Le plat pays*' – 'The Flat Country' – the other day, that really puts the power of surfaces and their angles (or flatness) into a dynamic perspective. There's this wonderful line (which I give in Tracy's translation): 'With a sky so low that a canal lost its way'. This is describing Belgium. Brilliant. The refrain in the song (poem!) goes 'The flat country that is mine'. I have lived a long time on lightly hilly ground at the bottom of a small mountain (or very large hill), with paddocks of stubble stretching out. And now at Jam Tree Gully I am starting to replant a rocky hillside, and to write poems in that space. The differentials, the segment of the line, discovering the nature of the slope, are part of the prosody of the poem's 'rise over the run': $m = \Delta y / \Delta x$ (see *Wiki* on slopes). I have always been interested in maths, and it's always part of my poetry – sometimes hidden away, sometimes overtly. It's certainly part of the poetics of gradients, of slope, that I am formulating now. By way of illustration I might copy in a handwritten journal page and include a poem....

> *30/1/2009, Jam Tree Gully, 9.45pm*
>
> Our first night staying over.... Kids are asleep and Tracy is in the lounge room reading (Stendhal). Looking forward to it all being ready for full-time habitation. Lot of work!
>
> Went for an evening walk with Tim and a pair of eagles flew up through the valley and over, then perched on large granite boulders – maybe twenty feet over our heads! Incredible. Walking

> up the steep slope the work on my heart and lungs created the waste, the chemicals, the energy of the poem. A usage of the excess my body produced in providing the work needed to get to the top, to the house. A poetics of gradients, of the hill. You see differently as you walk a slope, you sense differently. The angles of seeing change and you compensate, as does the poem. Even descent brings a work and a caution, the brain creates shorter lines to prevent slipping.

And here's some poetic work done a few weeks prior:

First Lines Typed at Jam Tree Gully

To hold the walls of valley
downthrust limbs of York gum
liminality, flakes of granite
and lichen scored as sun inland,
glitterati, this Toodyay stone
broken where the building
has opened precipice,
erodability, that movement
where we walk, dislocating
weight of conversation, even
meditation, to contravene
our visibility, perched
up on high, sidereal.

A drawing out, the day
lessens, rampage
of dead and living trees,
entire collapsed structures,
signs of fire as jam-tree bark
blackened crumbles with touch,
all working shadows thin up
the hill, the hill. Kangaroos
stir from their shady places –
the heat so intense at midday
they don't do more than lift
their head as you approach.

In the dirt, laterite smudgings,
hard-baked patches of sand, coarse-
grained breakdown of quartzite
in its granoblastic glory, a sheen
of mica and feldspar configuring
a sandstone past, a declaration
of origins; what grows in what
was here before? It demands
reconnection or the hill
will despoil to its granite

core and nothing more,
nothing more. Dazzling
anomaly of pyrites, breeze
sharpened with 'fool', 'fool'...
welcome here ... don't cling
together, give us roots
to nest among, cling to.

Common bronzewings heavy across the blank
of an arena we will fill with trees: sandy
spectacle, where horses rounded
on their tails: I see them twitch.

Internal fences down and out. Fewer
divisions. To predict a fate, changes
sweeping over an old old place; ring-
neck parrot feathers no divination.
What has chiacked in place
of undergrowth?

Weebills are here! And mistletoe birds
have been where mistletoe fruits have prospered,
have seeded the jam trees, where nectar-hungry
birds of many varieties test the hardy flowers
drooping in swatches from thin, straining necks;
the parasitic engenders its own chains of being.
I am not asking to be part of it. With time,
something will click, I have no idea what. No
second-guessing, despite the weight
of hexagrams, I-Ching. What else
I might read. Weebills are here!

Horseshoes and sheep skulls strewn across the block.
rare new growth, so late and odd. Fire wardens
watching afraid of vegetation? They have their own
version of prehistory, their own version of growth.

The making of place as a dynamic of couplings,
as if love and trust are omens, odds in your favour.
The sun burns but also fringes the leaves.

See Figures 12, 13, 14.

From the Land of Plenty ...

Figurative language can be a con – a way of having it both ways. I don't mean an ambiguity that suits a fence-sitter, but both ways in terms of 'artistic integrity' and making a political stand about an issue. As if the two are mutually exclusive. There needs to be a para-figurative language that denotes specificity without ambiguity (which I usually aim for in a poem!) in terms of tone and aim of critique, but that also allows for a flexibility in range of 'message', effect, and, indeed, affect. Somewhere between agitprop and the tone poem.

Here's an example. The poem below is an out-and-out attack on GM crops which the latest government (a right-wing government) has imposed on the people of Western Australia [the reference is to Colin Barnett's Liberal–National Party coalition, elected in 2008, which is certainly 'right wing' in an officially 'conservative' sense, but perhaps not by marked comparison to a succession of recent Labor governments in WA or to the present federal Labor government – NL]. A ramping up of larger-scale GM canola trials. They've been going on small-scale around here for years. GM culture has infiltrated in so many ways. Lies about 'feeding the world', when in fact the GM companies are lining their own pockets, don't wash, I'm afraid. That GM crops contaminate non-GM crops, and the non-GM environment at large, is well attested.

I have always found it somewhat ironic that some scientists accuse the anti-GM activists of defending organics as an act of class sabotage – claiming that organic food is a trendy middle-class fad irrelevant to the real world of commercial and mass cropping to feed 'the masses'. I feel everyone has the right to clean, unadulterated and non-GM food. The ironies multiply if one considers the middle-class privilege and appendages (cars, televisions, playrooms full of fads, conference dinners...) enjoyed by so many of these scientists themselves. Actually, scientists often consider themselves a class apart. What's more, the damage inflicted by their monocultures is visible and obvious. Something isn't working. Greed, government and corporate power, profit-taking even outside the Gordon Gekko realms, is part of it.

Gene patenting isn't altruism; it's research grants, jobs, kudos, flat-screen televisions and snappy four-wheel drives for inner-city driving (sorry, plus the odd drive out to check field trials). I rant because I want to indicate the passion behind writing the following poem. I mean business! I am being critical. However, I do believe in exchange and conversation, and have conversed with many a GM scientist about these issues. Of course, they can send plenty of criticism back my way as well, and I take that on board. So, though I am certain in my opposition and criticism, for this opposition to work in poetic form, it must let the language ironise somewhat my own certainties. To 'lose' the poem in vagaries would be against the aim, but to damn and leave no room for figurative language to take the reader elsewhere, to allow an escape or points of entry that are about their own way of seeing – I might as well not bother with using poetry at all.

Thus, I have written this in what I am calling a para-figurative mode (a poor coinage, but a pragmatic one), even if it tilts closer to the agitprop than the metaphoric. Main thing is, the figurative (however slight here!) works like a sheet anchor, dragging the bottom rather than holding the boat firmly in place. The ownership of 'plenty' is one that allows me to use this technology (the Net) in the comfort of a private (if shared) dwelling, with land around me that state law says is protected from trespass. Hence the 'para'. It's not a desirable state, but a critique of the certainties of agit-prop. I have plenty, compared to many. I still believe in the dissolution of the State, in the absence of government, in the value of community consensus, in veganism, in pacifism, in a non-GM organic world, in the conservation of flora and fauna and so on. But I have plenty, and that needs challenging. (I will add a second poem beneath this to illustrate a more figurative version of the 'para-figurative' political poem.)

Plenty

In anticipation of the moratorium
being lifted, farmers and their families
(because it's school holidays),
fly east to witness the first GM canola crops
planted for commercial gain in Australia.
The families follow the GM farmer (and the farmer
of GM) and snack in tents on not-yet-GM food.
This is the science that will feed the world.
And this is our 'proficy'. It's technology,
like your kids' Playstations
or the GPS in your four-wheel drive.
Food technology is stimulating.
The company supplies the seed
and a compatible herbicide; soon they'll meet
our needs for frost and even drought-
resistant strains. A company for the times.
Canola will thrive in dust. We'll feed the world
(if they can pay). It's scare tactics
when they rave about grain alcohol fuels
leaving Haitians feeding their kids
cookies baked out of mud and salt.
Enjoy the new science of the new
Western World: internationalist,
profit spread through 'starving
countries' where GM companies
have subsidiaries, outlets, local
scientists, shining offices, advertising
agencies, military affiliations,
farmers willing to feed us all. Back

in the West, spread the word,
let the kids brag about it at school,
topdress, lead the people to plenty.

Malherbe's Genetic Modifications: A Classicist Doctrine

Hey, I just heard the CEO of Monsanto
make odes out of genetic modification:
in steel rooms new crops make new promises,
their rules unclear, though we know
what he means, making doctrine
out of old-school plants, growth
pushed to the cultivatable limits.
Classical rules arise out of what's planted.
Farmers feeding profits,
engineers of rhyme and reason,

seeds and flowers inspiration
without imagination. Same locally,
though subtextual: wheedling
away in the background, test sites
embedded in crops, ribboned
and star-picketed. In the East,
the moratorium has been lifted,
and the West itches to follow suit.
Hunger doesn't have eternity,
and the body is a laboratory.

Clear as a world without imagery,
yellow canola fields effuse bienséance.
Direction as generic as pollen,
as seed in seed banks, as tailor-
made poisons in realms of poison
and lip-service; if you examine
the lubricity of canola seed, traipse
from one paddock to another,
it spreads its signature,
its overconfidence.

More on a Poetics of Gradients

I have been filling my journal – my hand-written journal! – with notes about writing poetry and living on a steep hillside block. The physical effort to move around the block is rewarding but exacting, even if one is reasonably fit and healthy. The string of poems arising out of this experience, this interaction (how could there not be poems?) – angles are what we deal with every day – is becoming central to the manuscript I am working on. Not all would enjoy walking 'our' place – some would come and look and leave before they had to negotiate the steepness, even with the many trees and rocks to assist in their movements. The *landscape* I am writing is in counterpoint to the landscape of Thoreau's *Walden*.

So, I have already written and published a number of poems about this, but I have also been keeping an ear and eye open for poems about climbing hills. For hill, valley, and high places, early (in particular) R.S. Thomas is superb. I am deeply attracted to his negative affirmations, his 'grim' negotiations with loss, and his beyond ironic observations of the imposition and liberations of patterning, mapping, demarcations (of nation, myth, labour, failures of belief and refusals to conform to imposed 'belief'). This, for example, from his 'The Welsh Hill Country' (in *Collected Poems 1945–1990*):

> There's a man still farming at Ty'n-y-Fawnog,
> Contributing grimly to the accepted pattern.
> The embryo music dead in his throat.

The Reverend Thomas's centralising of the human within 'creation' is always displaced (certainly in the earlier verse) by a discomfort with the relationship between human, nature, and creator, and often a questioning of what divisions exist between these. The hills themselves are an interesting case in point, almost existing outside the triangle, operating at a tangent (and gradients) to the will of things.

This morning I came across another poem that seemed apt in many ways. It's from a book I just received for review from the *Sydney Morning Herald – The Other Way Out: New Poems* by Bronwyn Lea. The poem entitled 'Red Hill' combines a clarity of affirmation offset by a haunting sense of threat that comes no matter the familiarity of climbing the same hill over a long period of time. It's a poem that has the language of gradient (though it doesn't specifically use that more mathematical descriptor/definition), poised to a point of absolute concision:

> – the acute
> angle of the world
> to my cheek
> rising as if to slap or kiss me
> even to lie
> down I am near
> vertical & filled with steep
> inclination –

The physicality of both body and spirit and the intellectual processing of the co-ordinates are deftly handled.

The hill *we* live on would not 'slap or kiss me' – it is too stony and too 'defiant' to register me or either of us in that way. Strangely, I know no matter how long I walk its slopes, it will resist me, but do so with indifference.

Gradient poems will always deploy words like 'steep', 'incline', and terms related to angles ('acute', 'obtuse', 'right'...), but the point is how these terms relate to the emotional, social, political, and ethical space of the poem. The specificity matters and brings different impressions. An acute angle makes for a very different tension from an obtuse angle: not only in the different description of place, but also, obviously, in the effect the angle has on how the reader 'feels' the description. You can climb a steep angle but you have to scale a right angle, so to speak. I am interested in the inside of an angle as much as in the outside. The radian of an angle is part of the world I am examining: part of the sphere, part of the 360 degrees of a conceptual (and a literal) 'planet' I am trying to create around the specific place, around Jam Tree Gully, around this de-mapped 'Walden' (or *not* Walden).

What I am getting at, under the weirdness and irrelevancies, is that when we paint poetic portraits of place, we in fact create a very real mathematics of that space. In so many fantasy novels, writers include maps – and indeed it was the maps that originally attracted me so strongly to Tolkien's *Lord of the Rings* when I was at high school (I went on to read the book twenty-six times and to write my final Year Eleven school paper in literature for my teacher, Bill Green, about it) – as a validation of the 'fantasy' (and/or speculative fiction) writer's imagined world, of its '*real* spatiality', as if the map makes concrete the projected imagined.

But I want to do the opposite – in my maths I want to de-map, to undo the recording of topographical features and the routes between them by allowing those features and routes to appear as they 'naturally' do on any given occasion: to alter and morph with erosion, changing movements of animals, new growth and the death of old trees, and so on.

I have mentioned my problem with people disturbing the bush by using GPS technology to go off-track, to make new routes between points. I am not advocating the arbitrary making of paths through spaces, but rather the following of paths as they grow and change. A lot of what I walk along are kangaroo paths. No map can be 'real' outside the moment of its making. It needs constant adjustment. Google Earth is an imperialism of mapping, not a liberation from its strictures, in the same way as GPS denies what's there to follow points given without subjectivity via a satellite. How the satellite looks is relevant: whether it's speaking electronically to a GPS or we're examining pictures taken. The use of angles, the use of circles, the use of poems written out of the line (out of Eudemus), out of rays, is an alternative to merely creating a 'map'. It is active and changing. Adjustments can be made.

Maybe pertinently, a couple of days ago we took young Tim to the Maritime Museum in Fremantle. Actually, we took him to see the Bon Scott statue, but after that had some spare time so thought we'd take him to see the harbour. One of the exhibitions was based on French exploration of Australia, and the display of early maps was fascinating. Curatorial paraphernalia confirming the drives of 'exploration' rang loud, and the stuffed animals spoke their own refutations, but the early maps of the southern land with their mix of hypothesis, guesswork, and solid charting, affirmed for me that the Idea (Platonic Forms, in some ways) of place is more accurate than the material. The maps, the more confirmed they are, become imitations of the land itself, and through imitation become separated from cause and effect. In other words, the map, though made through the senses, actually desensitises a response to place.

Over the last week the locality around Jam Tree Gully has been assaulted by scramble and quad bikes hacking up the reserve firebreaks and along the boundaries of where we live. In swathes of ecological destructiveness, these violent pleasurists (along with the shooters around the place) use intimidation and aggression to impose their sense of control and mapping on animals, plants and people who live here. I am told that many of them are city blokes come up during holidays to stay with other motorbike enthusiasts, getting overexcited at the 'space' and consequently 'cutting sick'. I understand their compulsions, and would be a hypocrite if I didn't acknowledge that as a kid around the farm I did similarly, but by the time I was in my mid teens I'd come to the conclusion that the destruction and violence of the activity cancelled the thrill and the pleasure.

Okay, so I will try and dialogue with them. Will the riders listen? Who knows? But that's the first activist step: dialogue.

Second and concurrent step: I will write poems, and maybe even read one to them. Seriously ... sometimes it works. Poetry can really matter. Many bike riders of whatever modus operandi are operating as 'outsiders'. I've known a few patch-wearing bikies in my time with whom I've had some excellent conversations about music and lyrics. In a territorial organisation, aesthetics and decodings are vital. I will certainly read poems to others around the place who are concerned about this damage (soil diseases and pathogens are also easily and readily spread via motorbike wheels). The terror and trauma being inflicted on the kangaroos and birds in the reserve, along with every other living thing, must be extreme, and needs to be countered. From the 'official' point of view, the riding of two-strokes through long grass is a fire risk, and this is another concern to all who live there.

But I mention this for another reason. One of the things that attracts riders to this place is the topography: the firebreaks cut through rocky ground, through treed slopes that become incredibly steep in places. It's a rough and no doubt exhilarating ride. The riders charge down one valley wall and loop up another. The roar of their bikes shatters the valley air with a ringing that is beyond any 'for whom the bell tolls'. The riders communicate with revs from one point to another. Triangles bend in 3D space. The rush down to the base of the valley (which is not broad) and the massive impetus downwards, then pulling up and slowing before impelling the machines with more revs to charge up the other side, is like a random and cut-loose amusement park ride. The down–up–down thrill. The bike does the work, but the rider's excitement and consequent bodily activity work in parallel, or in unison. In the act of thrill, the riders' is a poetics of gradients as well. Not about conservation (as I hope mine is), but about concentration, focus, and separation from real time, from the apparent limitations of ordinary movement (without the aid of a motorbike or other machinery). A kind of simulacrum of natural movement (the riding of a motorbike being more in contact with the elements than the driving of a car) that multiplies the adrenaline outcomes with less of the physical exhaustion.

The experience is repeated over and over until it sends those experiencing its side-effects around the twist (sorry, but it does). A machine poetics of gradients. If a bike travels up a hill at x speed and the hill has a gradient of...? It's their outlaw poetry, but what I want to tell them is that it's actually conservative and predictable: the outlaw act is, rather, to try to protect the ecology and to resist the urge of adrenaline, endorphins, and 'pleasurism' and 'leisurism', no matter how dark its origins, no matter how angry its rationalisations, or, conversely, no matter how much 'fun' it might be.

GPS and Forest Destruction

I was going to send a letter to the GPS-walk-people in the region, but will post it here instead. A public letter, as most of us are to some degree complicit, I am sure. I mention Kinsella Road in this letter, as its name is used on GPS-walking websites arbitrarily; not that this is anything like the way non-Indigenous people use town and other locality names that have been stolen or taken from Indigenous languages without second thought. In fact, they take ownership of them. In the language of co-ordinates, all signs are emptied out. 'Kinsella' as 'sign' has no more importance than any other, but my point is that all signs (and descriptions, be they a reference to a tree or a human-made object) signify something other than a means to an end.

Ironically, my Kinsella grandfather was a head state forester and the entire region of forest known as Gleneagle and beyond is still named Kinsella on state foresting maps. I don't need to explain what this signifies in terms of name overlays, implications of imperialism and exclusion of the original and ongoing Indigenous custodians/traditional owners. I use the word 'owners' in a general sense of relationship with land, not as a surveyed, material notion of property, which I reject no matter who claims land: for an anarchist, land ownership is nominal (I well know the situation), and arbitrary. It is a compliance to demands of State, but can be worked around by denial of its absolute value. Real 'ownership' is about caring and respect, and can be transient as well as continuous. But ultimately it's about knowledge and preservation, and GPS notions of land relationships appear to exclude these. Anyway, I refer to Kinsella Road because it fits within the GPS-walkers' own naming system and at least allows me a point of entry within their own GPS co-ordinates:

> you might notice my name is that of one of the roads used as a 'waypoint' where you gps WALK. it is named after my grandfather. i actively campaign against abuse of the forests – tramping dieback through the forests, lack of consideration for wildlife and flora and in general, are as horrendous as selective logging, horse riding, and other such exploitations of nature. people who require satellites to interact with 'nature' don't seem to me to have much empathy with the natural world. none of us can afford to be pleasure seekers who see the world as their playground and resource. it is hypocritical or naive at best. furthermore, the areas you cross through are often sacred to indigenous people and deserve more respect on this basis alone.
>
> reducing forests to gps co-ordinates, and the desire to go off the trails and destroy what little remains of the natural world, is rapacious and needs to be challenged in every way possible. i certainly will be campaigning against your activities.
>
> john kinsella

I propose a discourse of 'de-mapping' (best expression I can come up with at the moment!). Basically I want to de-map rather than map things. More on this later....

Left of the Left in the Arid Zone

Train across was uneventful, though I managed to get quite a bit of writing work done. I am still disturbed by my encounters and confrontations with shooters the weekend before. As the train passes a relatively short distance from Maralinga, I am reminded why one has no choice but a life of resistance. [Maralinga, home to the Tjarutja people in the far west of South Australia, was the site of 'top secret' British nuclear testing in the 1950s, undertaken with the consent of the Australian government; parts of the area remain contaminated, and compensation claims by the Tjarutja and the families of ex-servicemen 'poisoned' by the tests are still before the Australian courts. – NL] Travelling less these days, and by land and sea, I am reminded how protest is something best achieved locally. It's how you eat, how you shop (or don't), how you live and conduct your life. The most effective form of protest is surely to reduce the hypocrisies of your own terms of existence.

By midday Monday, I'll be a relatively short distance from Roxby Downs, one of the cores of the uranium mining industry in Australia. One of the factors I find defeating in cadre political gestures is that one is counted out of a protest equation if one is not on the line with others who are defining a moment of protest. Protest needs to be holistic, ongoing, and to be reflected in the life choices we make. Roxby Downs should be the site of an ongoing Greenham Common-type presence of protest, or people should entirely abandon any product, no matter how indirectly, that comes out of that mining source. Exporting uranium for the energy industry (aside from what gets consumed by the arms industry, whether admitted to or not) means that goods from software to computers to clothing and foodstuffs have been produced in countries utilising nuclear power as an energy source, and in purchasing them one is supporting such usage. It is the most indirect forms of consumption that are the most insidious.

My point about cadre protesting – the confirmation of one's presence is not enough when the material usages of those making the protest support (however obscurely) the things they're resisting. A means to an end? I am not convinced, and those who are least consuming the products of marketplace economies (always dictatorships, be it through so-called elected government or so-called single-party states or corporate interests) are those we don't hear from, unless their action is local, and they are acting on their home space, and we see them incidentally on the media. For those protesters doing as I am doing (and, I promise, will stop doing soon), and using a computer in any way, you are acting neither locally nor with impunity in terms of culpability. Your protest against environmental degradation collapses before it begins; the ravagers of environment hide behind their lies of sustainability and/or brute force, or legal and political sinecures – such protest is hobbled by its irony and feeds into the perpetrators' hands.

Mobile phones are the biggest irony of all. Consider what goes into the making of them, never mind the implications behind their use. Statements like this make those of the left you have associated with over decades, by way of agreeing to differ regarding tactics and ultimately aims (my aim is for small communities acting on consensus and with an ecological respect rather than larger centralised majority-rules outcomes) – seek to deny you or to separate you off from the left.

Where does that leave you? Not with the right, who try to shoot you for being part of the 'loony left'! As an anarchist, I am happy to be placed outside everything, though the socialist underpinnings of a strand of anarchism I have long been interested in necessitate my conversing in a spirit of co-operation with the socialist left. But I am as anti-Marxist as I am anti-fascist. Which is not to say I don't read and apply Marxist epistemologies to my reading and thinking (and writing), as I do, but doctrines of Marxism are as far from anarchism as anything else.

And it doesn't suffice for those of the left to separate those in university enclaves from those living in inner-city squats. A lot of us have 'been there'. And if you're there and starving and drug-addicted, you don't want to be there. (I do entirely support inner-city squatter culture, though.) Let me make it clear, universities are systems, and all systems need resisting and undoing. And yes, I do believe working from the inside out is effective. And I do have to help feed a family. I have reduced to half-time in order to plant trees, grow vegetables and change the dynamics of my life. Not as some 'life-change' scenario – I lived without any amenities and outside all 'acceptable' societal conditions for years on end when I was younger – but as a statement of refusal and also to reduce my impact on a very fragile and abused land.

Lightning Place – Jam Tree Gully

Yesterday was a stormy day throughout the district. When we arrived back at Jam Tree Gully, Tracy flicked the power on (we keep it off whenever possible and by October, as we've noted before, plan to be entirely without electricity other than some solar cells to use only when absolutely necessary), and found that no water was coming out of the taps.

This meant the pump had gone [in Australian, when something is broken it is said to have 'gone': short for 'gone on the blink' – NL] – we are not on mains water supply but rather have a massive concrete rainwater tank plus two air-pumped bores as back-up. I went out to check the pump, and found the housing around the power outlet that drives the electric pump entirely shattered.

A storm was in progress so I couldn't look closely, but we discovered today via Mum and J. (who went out for a visit) that it had been lightning-struck – with a six-inch-long and one-inch-deep scar in the concrete wall of the tank, the electrics burnt out, and concrete shrapnel everywhere.

The lightning strike had followed one of the oxidised hairline cracks through the thick concrete (reinforced with steel mesh inside) – a thin line of moisture that completed the circuit so to speak, allowed the arc across the gap to discharge.

I'd been working out in the storm, and given I've been struck by lightning in the past, and been in planes and houses struck by lightning on more than a few occasions, another chapter had been added (as J. said) to my memoir of 'lightning and me: a relationship'!

Random Access Clarifications

One hundred and seventy years ago today (i.e. on 23 May 1849), Thoreau, according to an entry in Raymond R. Borst's essential *The Thoreau Log: A Documentary Life of Henry David Thoreau 1817–1862*, 'Surveys land in Lincoln near Sandy Pond Road which he wishes Emerson to buy because of its beauty' (Marcia Moss cited in Borst, *The Thoreau Log*, 7). Although he probably means 'surveys' in a more general sense, the subtext is still of division and possession. This issue of surveying has always bothered me, and I tried to critique it in my poem 'Figures in a Paddock' back in the late

'90s (see my *Visitants*). I bring it up now because of the matter of removing fences and opening boundaries for the movement of wildlife (and, for that matter, people – but with the caveat: only if they're moving through to interact with place without damaging it – that's where questions of preservation become complex, and caveats make for contradictions).

Yesterday, BHP announced the first uranium mine in Western Australia. This is the beginning of another end. I want to declare loud and clear that not only will I protest this in my poems, but I will be seated where they are going to mine, speaking my poems as they cart me off. Land and rights, and permissions and access, are matters not only of consensus (of which there can be none), but of the long term rights of all peoples. Because the corporate State has constructed a set of conditions by which people have to rely on its largesse for basic human requirements, the need to profit from such mining activities becomes normative and seemingly necessary. If the land wasn't under pressure in that way, and community choices could be made without the imposition of a 'you will do this or lose' (at best) scenario, the State/corporate conditions nexus would be less likely to succeed in its tyranny.

Let's not for a moment believe mining companies and their government apologists are operating for the well-being of communities (local or otherwise). The entire dynamic of money, employment, security, rights, and well-being is a ploy to control: create the necessity in order to offer ways of fulfilling it.

As for 'progress', when we have another benchmark for this in whichever field – tomorrow, next year, a decade from now – few would envisage turning back the clock to today. It's circular logic that will be deployed against the neo-Luddites whenever the opportunity arises.

At this moment I am watching a female western spinebill doing somersaults outside the window. Its curved beak is an entirely adequate and all-encompassing technology. The irony of typing on this laptop as I prepare to go off-line and off-computer: well, it's a log book of a planned and permanent movement to find better 'technologies' (by which I mean less sophisticated and less reliant on industry: in other words, 'simple living' alternatives). They are 'pre', they are outside the notion of 'progress', and their usage is part of a desire to 'de-technologise', but yes, essentially they *are* technologies in themselves. Yet that's semantic, because what I am clearly trying to do is step away from material 'progress' and to say one reaches a point materially that is more than adequate; that in fact the damage done far outweighs the 'human application' regarding the ecological. A manual typewriter rather than a computer (a technology that doesn't need to 'develop' to achieve the same end results), a pen or pencil more often than not. Paper made from non-tree sources. And so on....

The 'planning': I make a living from writing, and have become computer-reliant in meeting my deadlines, as well as obligations and expectations of how text should be presented. I need to change the culture of production and how my publishers and others are willing to accept material from me. It can be done, but it has to be carefully planned and discussed. October is the deadline I'm aiming for in terms of 'home' stuff, with my university communications following at some point when I've been able to lobby effectively for some changes regarding my communication with students and so on. That can be realised, I am sure.

Actually, it's more than this: I believe that we have to rethink social notions of what is adequate and what pleasure and leisure are. But this is not 'primitivist' thinking: it's poetic thinking. Poems, to my mind, are about repair, analysis, fruition, and not destruction. And 'destructive' poems consciously deployed bring attention, in the cases I respect, to the failure of acts of repair. I write a lot about death and destruction, but I hope this allows a reader to refocus on their role (and the poet's role and the poem's role) in making such things allowable.

I had an interesting exchange with a fellow poet (and one I admire) the other day, about the blurring of lines between activism and poetry. He felt they were separate acts and used Judith Wright as an example. I maintained that Wright was an activist in her poetry as much as in her general life, especially towards the later years of her poetry writing (clearly in her prose, she was). This poet friend was talking over his reactions to my *Divine Comedy: Journeys Through a Regional Geography*, which I consider my main work of poetry, a work that is in the realm of the 'parafigurative', where activism and poetry go hand in hand. Not to be didactic, but to be suggestive, and prompt 'action'. My fellow poet felt that one should choose either a life of activism or a life of poetry. My reply was as follows:

> mum was a poet and I've been writing it since i was six. for me it was a poet becoming an activist and making the two talk with each other. i have a book on 'activism and the poet' out with liverpool uni press next year. I'd like to think my poems do something other than 'tell' – i try to create many levels of approach in every poem i write, and for every poem to be reinvented with every reader and every reading. i am writing a new intro for j. wright's selected at the moment and think she was an activist-poet in whose work (later work especially) these elements were in synch and didn't counter each other. poetry has a long history of activism. on our blog (mutually said) i use a coinage i came up with re metaphor and activism – the 'parafigurative'. this is what i am trying to do – articulate a poetry of action but also ambiguity.

My poet friend also went on to discuss the structure of my *Divine Comedy*, and as this is relevant to the reading of the poem in terms of local and regional activism, I'll include my reply:

i think there are four narrative threads in the book:

1. the template of dante
2. the movement around the block approached from different angles (per the different canticles)
3. a topology and taxonomy of place that builds and 'collapses' to rebuild which is intertwined with a literal history of the place – including the building of a portrait of a surrounding community (yes, williams is the right parallel in this sense, and even more so olson's maximus re location and illustration by example and observation and snippets of history etc).
4. the interactions of the 'characters' involved re their epiphanies and 'elegies'

Now, it's back to my Thoreau book. Have just completed two new poems. Am now in the process of developing a 'narrative' framework for the book as a whole. It's a matter of reconciling surveyed areas of poems and the points of access that surround them. I guess this will make sense (I hope) when the book appears. At the moment, I am considering 'random access' versus a set of semantic points of entry.

Forest Sculpture Walk Seems Anti-Forest To Me, Sorry...

Just saw a short on SBS Television featuring the Northcliffe 'Understorey' forest project (in the south-west of Western Australia). I was the initially successful applicant for the position of poet-writing-the-forest and pulled out due to my distress at the mistreatment of the forest involved. Bulldozing a chunk to make an arts centre, hacking a path through die-back sections of the forest. I went down for a meeting with the Northcliffe organisers and though some considered my concerns, they rejected my requests for dieback treatment areas and so on.

While in the town collating material for my poems to feature in the forest, I interviewed some more radical members of the community and heard of their concerns regarding the forest being used for an arts exercise with the forest coming second and the 'artsy' stuff coming first. Various ecological concerns were voiced, and I realised that to contribute to the forest project would be to violate everything I believe in. Rare parrots would be disturbed, banksia dieback (banksia is a sentinel species for die-back) would spread in the making of the path (or boardwalk as it turned out to be), and the intactness of this small piece of forest would be disturbed – carved up by the walk and the movement of people.

The best kind of sculpture in the forest environment/ecology is surely 'created'/'evolved' without human intervention and is chanced upon (if 'discovered' at all) and left alone: not constructed and revisited treating the forest as if it's a 'creation' for humans to use as an art gallery (basically acts of mimesis, anyway). This is the triumph of aesthetics over nature, and the ecology that is appended to make for good consciences is a lie. The primary local argument for this was that it was better than logging it – this is true, but it should be added that the piece of forest concerned was a gift to the town and basically outside logging jurisdiction. To leave it intact and to keep invasive 'artsy' hands off it would seem altogether the right thing to do.

Anyway, my withdrawal meant other writers filled the space and provided poems anyway. There are many who would disagree with my position here, pointing out that at least some kind of preservation and understanding of the forest is being expressed. Maybe, but I feel that when the 'arts' are served first, the forest will always come second. I don't doubt at all the good intent of the participants, but I do doubt the efficacy of the endeavour with regard to the ultimate health of the forest and the creatures that inhabit it. It's a question all arts practitioners have to ask themselves: I guess many would see my view as both self-defeating and unrealistic.

CHAPTER 12

Poetry, Justice & the Court

[For Derrida, there was perhaps only one question that ever interested him – the question of justice. Justice, for him, always remains to come; it is simply not possible, therefore, to expect justice to be done on the basis of courtroom decisions and legal precedents, and yet this is no cause – far from it – to give up on justice (see Derrida, 'Force of Law', and Lucy, *A Derrida Dictionary*, 62–65). But while the question of justice could be said to have been an abiding concern of Derrida's, it is not a question he always addressed directly. Similarly, we might see that John's interest in the question of justice is an important, if often 'subterranean', theme throughout this book. Here, he responds to that question openly, by way of its relations – of course – to poetry. – NL]

I

The public space of a court is a crossover zone where the real after-effects (and pre-effects) of 'Justice' are felt. Courts are active liminal zones. There's a poem by Nyungar poet Lloyd Riley, entitled 'Supreme Court Gardens', that encapsulates a lot of the contradictions between the loss of one law and its usurpation by another, the occupation/colonisation of physical space, and the codes of the land being covered and substituted by new codes (urban gardens). The gardens of a court (and its public spaces in general), or those in its locale or bearing its name, become complicit in a greater dialogue on which law is the right law, and whether the mere presence of building and landscaping denotes authority (it doesn't). This poem (I will reproduce only a few stanzas) shows the power of the dreaming, the power of the imagination to transcend the urban, the building over the traditional, to literally reconnect with what is still 'deep, deep' down (by extension, the traditional law and way of living):

> What do dreams do? Is it a twilight zone?
> Is there a doorway we all pass through,
> Not knowing what's on the other side?
> One Nyoongah came back then disappeared.
>
> He's in the other world, not dead!
> Living here with urbanisation, yet
> Living faraway, deep, deep down,
> Down behind those mournful eyes,
> And above that empty laugh.
> Above the gentle blue skies
> Below the concrete and the tar
> Living beyond the touch of 'progress'
> And away from the snarling jaws of
> Civilisation.
>
> (In Brewster, O'Neill and Van Den Berg, eds, *Those Who Remain Will Always Remember*)

It seems essential to me that if there is to be any reconciliation in Australia between Indigenous peoples and migrant Australians (all other Australians), it needs to start with the court buildings themselves, and, obviously, the law. The oral and written poetry of Indigenous Australian poets is constantly concerned with law, and justice, and should be incorporated into *all* legal buildings in consultation with the traditional owners of the land.

II

It would not be difficult to render court language into poetry, or vice versa for that matter. So much of the court's duty is in listing and collating detail about specific crimes – the type of objects stolen, for example, their value, the specifics of items (if a shoe, it might be size, materials used in its manufacture, type of sole, etc.), the brand names and the import of these, and so on. The accumulation of evidence is vital to the presentation of a case, but though a trial might turn on a very specific piece of evidence, the overall issues of wrong and right are broader, more generic. It's a mixture of the exact and the subjective. So too with a poem – some poems work entirely through the accumulation of close-up detail, others speak in clichés and generalisations. When we've read a poem we usually get a sense of a generic theme or mood: death, life, love, conflict, harmony, and so on.

Terms of address and ways of listing are overtly poetic. Whether law from the Bible or Koran or other religious text, the secular variation isn't often that far away. The secular court still has its roots in the religious. For some, the de-spiritualising of the court would result in the court's non-recognition. Poetry is a way of bridging this gap. In my edition of the

Koran, the commentary points out that the majesty of the book's poetry (even when it is laying down the law with great specificity) led to whole eras when the poetry of the Koran, because it was so effective, became the poetry of entire communities, meeting literal and figurative needs. I am not suggesting secular courts should become religious (I am an anarchist, remember), but I am suggesting there's room for dialogue there.

While much has been written on the relationship between poets and the law, there has not been a great deal written about the inclusion of poems and poetic texts in the design and building of court spaces. As a subscriber to 'natural law' – and one often politically sceptical of the legislative and executive motives for implementations of particular laws, and the nature of potential punishments attached to applying those laws – I feel it desirable for there to be, in effect, a complete absence of legal machinery. However, I am in many ways a pragmatist and feel that rather than functioning as vehicles for depriving individuals and groups of freedom and liberty for their transgressions, courts might serve as places of dialogue on the nature of these transgressions, and the appropriate way of dealing with them. For the purposes of this essay, I am not pursuing issues of the rule of law and separation of powers; suffice it to say that I am arguing for a 'metaphoric' or 'figurative' dialogue, rather than a legal, logical, 'realistic' or specific one.

I would argue that the language of courts, seeking to be the most precise form of language, is in fact one of the most poetic. When the layperson in court gets lost in legalese, it's not simply a case of losing track of the logic as clause breaks into sub-clause, but that, heard as a whole, it becomes a blur – too much information to follow without close study. Mishearings and misprisions create a poetic distraction, a figurative language that moves to its own rhythms, creating a metaphoric slippage in which missing a disclaimer or caveat might lead one to interpret the text 'against' its intentions. This is a crude reduction, but the point about loss of meaning is a common complaint. Of course, a piece of law that has been tested in different contexts is not necessarily fixed, and precedent cases or challenges mean that language evolves, is added to, even deleted. This compounds the confusion – a misunderstanding regarding the organic nature of textual (written) law in particular.

If I might extemporise: a court in which there is never doubt is a dangerous court, a court that has lost its humanity. Taking it one step further, though error is the least desirable outcome of the court (the executive – and maybe elements of the judiciary itself – might well be happy with any error that suits its purpose), the error that veers towards caution in condemnation (someone let off for a crime they have committed) is an 'intimated' and humanised – and metaphoric – gesture.

III

In his 'Writing, Law, and Legal Practice in the Athenian Courts', David Cohen quotes Demosthenes from *Against Meidias*:

> And what is the strength of the laws? If one of you is wronged and cries aloud will the laws run up and stand at his side to defend him? No. They are written texts (grammata gegrammena) and incapable of such action. In what, then, resides their power? In you, if you support them and make them effective whenever anyone asks. So the laws are strong through you and you through the laws. (79)

At the beginning of his essay Cohen differentiates between the written and 'unwritten, unchanging, timeless divine laws', and here he emphasises 'the mutable, man-made, instrumental quality of law'.

This is an issue of prose and poetry for me. As an anarchist, I am not against a rule of law – I differ on how laws are made, why they are made, and how they are applied, but am not against their existence. I work small-scale rather than large-scale, and feel that most people know the difference between the right way to behave towards each other (and the world at large) and the wrong way. Now, that's a subjective issue of choice, but it's in Demosthenes that there's an argument to be made regarding language: 'So the laws are strong through you and you through the laws'. True, but not if the meaning of the (written) laws escapes you. Not if you have had no training or experience in understanding them. *Is* this a case for simplicity: fewer laws expressed in the simplest language? Clearly impossible, at least in the context of the modern nation state. Then what? The space where poetry and prose blur: people might better understand through the antithetical, the ambiguous, the metaphoric. In a practical sense, discuss the language of law differently – overtly introduce poetry into the court buildings. Poetry and prose as one.

IV

I'd like to look at three examples of poetry I have written for use in courts. Those on trial have not infrequently read poems aloud in courts, and that is another function of poetry, but I am talking specifically of the poet writing from 'outside' (though I have also been 'inside'!), for decorative (if political) purpose. As installation *and* intervention. The poems were written 'about' the space of the court, but also designed to be displayed there. In the first case, this became a reality after a judge read the poem in court; in the second I wrote it specifically for that purpose. In the third case, I wrote a series of poems to encourage people to stop an execution, and having failed to do this, would like to see the poem built into the fabric of court buildings as a reminder of the negations, contradictions, and implications of the death sentence.

In each of the poems I have written for courts, devices from metaphor to synecdoche are at work, as well as the genre blurring and 'message'. The 'mixture' of technical approach, and the choice between open and closed verse forms are vital: the tensions and relaxations that dialogue within the poems seek to replicate something of the to-and-fro of presentation, cross-examination, and summing-up of the courts themselves.

The poems have come about circumstantially on one level, and by overt intent on another. I have, in the last few years, written two pieces for Federal Court spaces – in Perth and Melbourne. Both are in 'open' forms. The first poem, the Perth Federal Court poem, is about what makes law, about its social implications, about the loss or replacement of others' law, and about obligations to protect not only the rights of people, but also the rights of the environment, animals, and place. The Swan River, which flows through Perth, is used as a metaphor and symbol of law and connection. It is the thread that makes sense of cause and effect: of decisions and their consequences. It was written looking out from a place of dialogue, hoping to extend that dialogue. It was written looking out from a place of restitution and hope, and a place where hope might equally be lost or destroyed. The river suffers a similar fate. It calls for a holism and a respect for those who lose as much as those who gain. The use of 'pollution' is heroic Greek as much as literal.

The Swan River Viewed from a Court Balcony

What flow passes the vista
of a federal court,
pauses in surface tension,
the band of green
that makes a foreshore?
Blocked out, top-dressed
for rally drivers,
but still an official airstrip,
that might be redesigned
to influence the light – growth,
sculpture, running water,
to give back when the fad's
to take, to grow out
into what now amounts
to a lake connected to a river
at either end; from this balcony garden,
this judge's bench floating
or squatting over the courts,
that cast down or lift up
or just ask to be heard,
a social conscience
that might be the flow of water,
of information, of language kept

in check and saying exactly what it means.
It's a framework too, it's a molar
composition, the molecules settling
against the banks, swarming
about the jetties. The spray,
the drifts, the currents and tides
inform a judgement: the random path
of particles cannot be forgotten:
we're in there, like it or not,
breathing the same air,
the same water.
It's pollution
we want to keep out,
but metaphors and symbols
don't always bridge the gap: people
are not algae, and the blooms
that choke the river
find no place in the language of beauty:
the liminal can be exact,
and when we forget that
people are wrongly imprisoned,
and justice wallows in the drying
riverbed. A kingfish moves
rapidly upriver, avoiding bait
and pollutants; it sets a precedent,
as if it co-exists with the city,
the river's traffic, the language
obscured by development,
any number of histories.
We might imagine
its arguments.
Summing up…
                    the water forms
a painting: as temporal as we
make it, its laws mouthed
through sediment. Curators,
we sail against the light
and forget, until something
makes us look back.
From up here, the river sits,
its stilled points
holes in the text.

The second poem I wrote from inside out – from a Federal Court building looking out – was in Melbourne, in 2004. The Swan River poem was dedicated to Justice Nicholson, with whom I have enjoyed many long conversations about the nature of law and justice, and with whose hospitality I was allowed onto the court's balcony. In the second case, accompanied by Justice Nicholson, I was introduced to Chief Justice Black, who kindly showed me his chambers and the court in general. I was excited by its poetics: in general design, but also in the use of text-in-glass. The poem I responded with is about the nature of the court itself: specifically the Australian Federal Court, and the nature of 'Australia'. It's also a poem that works with the discussion I had with Justice Black about the permanence (and impermanence) of law, and how and why notions of justice might change. The stone of the building becomes the obvious symbol of longevity, reliability, even permanence, but it's the view that changes, the lines of the map that do liminal double-duty as permanent yet constantly changing (even where change is slow, as in a desert landscape looking a long, long way outwards). For me, it was the light of the building that gave hope for dialogue: as in jails, light should never be denied. It is as important as food or shelter. To deny it is a crime as great as any. In this new building, light reached into and around chambers, points of viewing in and out were provided, even the courtrooms experienced the enlightenments of angled and abundant light. The city (and the gardens/park over the road) became part of the interiority of the space/s. The court becomes a place of liberty, release, and thus justice. Hope is, in design at least, fed, despite the contradictions of colonisation, occupation, and displacement that a court building will inevitably also represent and symbolise. It's the recognition that justice is positioned looking in and out of the pragmatic space as much as any other.

View Turned Inside Out from the Federal Court in Melbourne

for Chief Justice Black

You say 'up the Centre to the Gulf,
that's where Burke and Wills went',
and our view augments the possibility
of the mind's eye taking us away –
the stone these laws are set in
changes with heat and cold,
with the lie of the land,
the flow of river
and the deepening clouds
of interior, intensity
of rainforests, textual ripples
of wheat crops, ochre and primary
against the red sky of stories,
parables – the echoes' timbre
a variety of stone, this building's
plateaus and chamber, edifices
and rills, tablelands and highlands,
caves and cliffs facing broad oceans,
climatic interface and translation; to add,
and 'across into Australia Felix',
the emanations of Yarra

weaving its suspending waters,
hoists, chains, trucks, forklifts shaking the docks,
Mount Macedon ... places to get your
bearings by, places you rely on.
Working the containers,
stevedores know as well these
fulcrums, and their gateways
opening out beyond satellite navigations,
the type of product they load and unload,
the routes the crow flies inland; to reassure
the sentinel balance, Mount Dandenong,
where my partner's people
congregate, out of Ireland
and the goldfields,
entranced by the lustre
of good fortune, a semi-vision
half-remembered in the wavering
light between us; this balcony
that holds us like justice
wrought from light, illuminating,
exposing those who'd
refuse to listen ... *we* crave
light and access. Threefold
the leaves of glass
filter light and history: legislature,
executive, judiciary ... the architect's
veil of words, or all we expect
looking up at edges
of letters testing opacity, processing
aspiration without tension
as if the flow of the river
were never in doubt,
its visibility
unravelling smoke-stacks
and ships, the bleakest cases....
The green light of the park infuses,
hints at early darkness ...
a chevron mirror
turns the court inside out,
the spinal cord so sensitive,
waterholes and campfires
marked in every seat,
faintly overheard
by those not listening
to the explorer's laws
of stone, bush, and *lineless* maps,
the songs of light
there before them,
tracks mapped out
between the camps.

Over the years I have written a number of 'protest' poems about state execution, to which I am deeply opposed. To be effective as poetry, however, these works utilise slippage and metaphor, and invite the reader (or listener) into dialogue or conversation with the poem. Most recently, I wrote a series of poems for Van Nguyen [an Australian citizen convicted of drug trafficking in Singapore, who was hanged – despite pleas for clemency by the Australian government, Amnesty International and others – at Changi prison in 2005 – NL], hoping to help stave off his execution. These poems were written to anyone who might listen, but most specifically they were written to the idea of the court in Singapore, and very much to the building itself: that is, where the law was interpreted and applied. The city becomes a metaphor for the court itself, which is not to hold all responsible for the outcome, but to remind us all that we all have a voice and place in the application and preservation of justice. I consider myself part of a system of justice, and culpable when justice is wrongly applied – wherever that might be, allowing for cultural sensitivities, but still maintaining the right to life for everyone, everywhere. The court building should be a passive defence against violence in any form, to my mind. When the violent go there to be tried, they should be prevented from making violence, and receiving it! If not, it's hypocrisy, I feel.

The form of a poem matters greatly to how it might be read. The following poems are written in strict traditional 'closed' forms (though I try to innovate within these parameters, as I wish to offer an alternative to the 'law' as it exists in the context), and those forms becomes cultural/intercultural connectors, signs of respect (across cultures), so as to attempt not to disrespect another people's laws and values: I will engage as much as I can on 'your' terms.

That is, not to create an Us and Them situation (in the example of the Van Nguyen case, many Singaporeans felt Australians had a purely nationalistic, primarily economic and even racist agenda in taking a sudden interest in this case, as opposed to others). At one stage, a journalist from the *Straits Times* in Singapore approached me to do a feature on the poems and campaign (days before the execution), but was vetoed by the editor. I sent the poems to the *Straits Times*, to *The Australian* newspaper (which responded only with silence), and to a Singaporean literary journal and discussion board, which did publish them instantly. My hope is that these poems find their way onto court building walls, as reminders of our obligations.

Grace: a plea for the life of Van Nguyen

The humidity is increasing here,
and stretched taut through a time zone
we share, there is no argument of nation,
but rather a simple equation
of what's done between sunrise

and sunset – our washing, our eating, our prayers.
In taking a breath that feeds the air
with embodiment, we gasp,
struggle to fill our lungs,
struggle to fill our days
awaiting the sleep we will wake from:
fresh, contrite, enlightened.

Doina: another plea for the life of Van Nguyen

The serrated fronds of palms waver at the airport;
The condemned is cut off from flight.

The ripe durian fruit is sweetest eaten when the smell's offensive;
The condemned eats, is killed, the bitter-sweet unbalances.

The flowering strands of rice exhale inflorescence;
The executioner eats a meal – before, and after.

Tradition: a third plea for the life of Van Nguyen

Even storm clouds of crushing weight
Bring life – brood and effervesce.
Even those who feel it's too late
For clemency, might reassess.

Singapore Orchids – a fourth plea for the life of Van Nguyen

The moon and sun
Have fused as one;
Tiger orchids –
Unseasonable.

Mirrors inflect
Inside out – flecked
Trees suddenly
Cool – bright perfume:

Pigeon Orchids.
And Kasem Gold
On Orchard Road.
Taste evening.

Light in the cell?
Growers desire
The exquisite
Seraman live

Longer. Mandai
Orchid Garden.
Botanic light.
Controversy:

In accepting
The Vanda Miss
Joaquim Orchid's
journey to light?

Singapore – your
Orchid... your light.

Lightning – a fifth plea for the life of Van Nguyen

We observe night places
so intensely
when rips
of lightning
underscore:
though we look
harder for it,
harder
at the ragged edges
of rock and foliage
levelling apostasy?
The storm ranging
through the valley,
still the smell
of cut oat-stalks:
so brittle
earlier in the day,
broken open
to fuel violence
everywhere
around us.

Preservation Haiku – sixth plea for the life of Van Nguyen

Elevated birds;
Plants photosynthesising;
No execution.

We celebrate birth
To enrich the biosphere;
No execution.

Eyes preserve distance;
Up close makes large of the small;
No execution.

The grime wasn't there;
Rain over the window pane;
No execution.

The boat cuts water;
Its wake is soon forgotten.
No execution.

The curved horizon;
The intactness of island.
No execution.

Ghazal – seventh plea for the life of Van Nguyen

Buoyed by rain in the volatile forest, it's our loss.
Forget the risk of fire and take your rest? It's our loss.

Children in my daughter's class will celebrate a birthday.
All consequences as they feather their nest. It's our loss.

She wants to go but feels she shouldn't. It's her conscience.
Whatever she does we'll support her. A test? It's our loss.

On the coastal plain it was raining less – people barely noticed.
Though noticing it wasn't as hot, felt blessed. It's our loss.

I watched the ornamental fish swim golden in soda-blue water.
Surfacing as one, I heard their request. It's our loss.

You'd stand alone, John, and make your statement?
All of us lose a fragment. The State confessed? It's our loss.

The Facts – eighth plea for the life of Van Nguyen

The science of execution
can be studied at night school,
soon it will have its place in the universities,
but with emphasis on the coronial,
forensic pathology of state's evidence
turned over and buried
in research hospitals,
on television.

Darkness – ninth plea for the life of Van Nguyen

I walked through darkness this evening
but it wasn't darkness at all –
there was no resistance,
no pushing back.
I imagined the river
flowing nearby – fiercely close
gatherer of tributaries.
I could sense the movement of fish
through oily water,
I could hear movement of light
on the bank distantly opposite –
but there was no resistance,
no pushing back;
it not being darkness at all.

Luc Bat – Plea for Van Nguyen on the eve of his execution

Each life contains colours
Of the spectrum, and soars over
The darkness – discover
This now, and recover the love
Lost to shadow, above
All else cherish the love nearby
Us – friends and family,
Mother's hand desperately touching
Van's, alone, though sensing
Us close by him, remembering.

CHAPTER 13

The School of Environmental Poetics & Creativity

Prefatory Note to 'Manifesto For a School of Environmental Poetics & Creativity'

The following manifesto was prepared early in 2005 while I was teaching poetry at Kenyon College, Ohio. It was emailed to a hundred people interested in teaching and ecological issues, especially though not exclusively to academics and poets. Others 'invited' to participate included farmers, environmental activists, editors, artists, radio producers and Indigenous spokespeople. The reaction was overwhelmingly supportive and not only did different universities offer spaces for the temporary 'school' but I/we were even offered an old building to work out of in Western Australia. A network of like-minded participants evolved overnight.

One of the projects that arose from this school-without-centre was an invitation from Gerard Greenway to do a special issue of *Angelaki* exploring issues coming out of the dialogues around the school. Over the last few years, the school has become more of a conceptual entity than a practical reality. However, I would like to think the kind of conversations going on in this ecological issue (see Fagan, Kinsella and Minter, *Ecopoetical Pedagogies*) might relate to some of the concerns of the school, though, in essence, outside its auspices and operating on a different tangent. For me, that's become the essence of the school that is not a school – to trigger and engender broader conversations.

The original manifesto is included here along with an addendum – or a disclaimer – I wrote in response to elements of the conversations in the *Angelaki* issue. While the editing of the issue was not part of school discussion, the 'school' is one possible entry-point to considering what's said. There's no belonging, there's no ownership, there are no rules of reading – rather, there's a modus of consideration and investigation.

Since 2005 I have changed my views relating to the 'school' considerably, but have not stopped believing in its ultimate goals. As I have decided on a 'neo-Luddite' approach to technology – to reduce reliance on electricity and utilise 'simple' or 'basic living' principles – I no longer see the Internet as a viable means of 'sustaining' (fraught word) an ecological pedagogy by such means. In fact, in October 2009, I will cease using computers outside my university work environment, and even there work to define alternative modes of communication with students (an overwhelming irony, of course, given the merely recent nature of Internet communications in any case). I feel that if such a 'school' were to work effectively, then it must be as separate entities working entirely through the local, with at most an international–regionalist awareness of discourse in other 'localities'. Communications at a less rapid pace won't create a local 'dark age', but rather increase respect for the immediate and inevitably, I hope, encourage all participants, regardless of ethnicity and perceptions of 'home', to respect traditional custodians of place and learn to listen to country itself. This is a form of 'de-schooling' on the one hand, and on the other also an affirmation of community as a repository of knowledge that is primary *and* secondary source at once. When Margaret McMillan writes in *Education Through the Imagination* (1904), 'many a child passes through his school life using only a minimum of his powers and expressing only a fraction of his personality' (cited in Curtis and Boultwood, *A Short History of Educational Ideas*, 510), I concur (the 'his' aside).

However, I think this can be applied across the board with all learning in terms of the ecology that surrounds us as we learn. Whether a totemic or a less visceral relationship exists with place, many tend to use a minimum of their 'powers' in connecting with an ecology, unless they have been brought up respecting that ecology, and understanding its necessity to their own physical and spiritual health. Degrees of separation – starting with schoolrooms themselves and extending through to the apprentice's workshop, the factory floor, the adult education centre, the university, the night class – disguise the impact of losing contact with 'natural' environments. This separation is at the core of gaps between cause and effect even in pedagogy that is sympathetic to ecology.

I seek to redress this by taking the classroom outdoors, where possible, or, if indoors, utilising excursions and field-trips around the immediate area (parks, roadside verges, through to bush expanses) as much as possible, and not just as an activity relating to suitability when doing a relevant 'subject'. Maths can be taught in this way as well. The thing is to participate in place without leaving damage – to follow tracks already laid (from animal tracks through to open areas). It's a pragmatic approach.

In essence, to my mind, manifestos must remain open-ended.

Manifesto For a School of Environmental Poetics & Creativity

concept

To create an alternative place of tertiary study concentrating on the arts and environmental/ethical issues. To be internationalist in scope, but regionalist in its immediate geography – that is, in tune with its physical surroundings and reflecting on the issues of its specific locations – the school will seek to further a practical application of study to benefit life in general. Pedagogically, it will both complement existing tertiary education, and exist entirely outside such structures. Criteria for success are the benefits brought by individual and group study to the greater goals of environmental preservation and conservation, non-violent and peaceful dialogue, the furthering of self-achievement through collective action.

The school itself would be spread around the world, but, as indicated, the particular environmental, 'social', and 'cultural' issues of each specific place of teaching would be integral to the 'language' of that teaching. The original aim is to secure a building in York, Western Australia, and to make use of Edith Cowan University's Landscape and Language Centre as a nexus between 'official' educational institutions. The ability of the new School of Environmental Poetics and Creativity to sustain interest and functionality will rest heavily on such arrangements. Further connections with colleges and universities in Britain and America are being developed, with prospects for interactions with others around the world being mapped out. The prime point is to move away from ideas of nations and borders, and think in terms of interactive communities of varying scales: the smallest being as relevant as the largest. Technology will enable these interactions, but technology is not in itself an 'aim' of the school: it's purely a tool towards a less exploited world, and not an end in itself. Crafts of paper-making and traditional printing, with 'enviro-friendly' non-exploitative inks and bindings, are examples of an alternative technology the school will eventually work towards.

It is the hope of the founders that the school will attract academics and practitioners to participate (out of enthusiasm at first), in ways however small, to give the school the academic legitimacy it will require to persist and bring genuine change to the way we apply learning to the 'real world'. The school is about theory and praxis. In some senses, it is a concept that allows interaction with our pre-existing groups and communities, to increase the flexibility of its teaching and facilities.

practicalities

This letter serves as an invitation to participate. What we need is for practitioners and teachers to seek to associate their relevant pre-existing courses to some small degree with the intent of the School of Environmental Poetics and Creativity. If it is agreed that we all share some basic beliefs, it would allow for students to participate (for free at first) within the greater fabric of the school, without detracting from their standard education. If they wished, it would give them an association with the project, and a place of dialogue (via the Web mainly – but later as places to visit, if they wish), and also a place of exhibition, publication, and display. The same applies to teachers – the same opportunities, interactions, and occasions for presentation and dialogue would exist. A Web discussion list, a website, and a series of environmental publications, would mark the early stages. As mentioned, the securing of a building in York (or a few rooms to start with!) is high on the agenda, but the hope is that we can also operate out of Edith Cowan University, Western Australia, and other offices. Soundings-out have already indicated that leading poets, artists, musicians, and theoreticians are willing to be part of this. In many ways, the teacher is also participating, and the division between teaching and research is broken down: the teacher's research and artistic endeavours or field activities become the lessons. The students learn from an early stage to input as teachers: by assisting in the preservation of, say, native bushland, or a recycling programme, or a public arts project, or a local radio programme, they feed skills back into the project. There's a holistic rather than a systemic basis to the school.

funding

This project is basically unfunded and will rely on participation and belief in the positive end results of preservation and mutual respect. It is cross-cultural, entirely ecumenical, non-classist, and environmentally oriented. It is activist and non-violent, with programmes already designed for activities as diverse as writing and vegan ethics, the materials of art, cross-cultural poetics, sexuality and landscape, and numerous others. We are welcoming input here. 'Basically unfunded' has the disclaimer that input by universities, even indirectly, into the project is associative participation, and notions of independent purity are flawed as a consequence. However, these are not associations of dependence, but of intent. We believe that such interactions strengthen 'just' causes and enrich educational opportunities across the board. The school is a dramatic alternative, but the basic tenets of education are still in place, and largely compatible with the principles of knowledge expounded by other schools. A difference would be that we are concerned primarily with an 'ethical' or responsible application of that knowledge within the school environment. Clearly and emphatically this is often the case with other schools too – but we are interested in the investigations of the ideas around this, and have the general health of the environment in mind when performing our artistic or scholarly activities. The school does not seek to produce workers for the State, but a desire among participants for a just and equitable future for all living things. Direct funding will be pursued with this in mind, and eventually an ethics committee will analyse and vet potential funding (as well as seek it), so that facilities can be established and wages can be paid to teachers. There will be costs to students working entirely within the programme, but this will be kept to a minimum. The principle of free education is tantamount, and our desired end result; but it will be a communal journey.

accreditation

This will happen within pre-existing courses per the usual means, but the aim is to have entirely independent 'on-site' courses that will be 'marked' within the framework of the school's goals. The awarding of degrees is an issue that will vary from country to country, and the process of validation is a time-consuming one – our belief is that those who wish to earn their official pieces of paper can do so by interacting with the school through pre-existing relationships with traditional places of education. We have no problem with that – we hope that it gives them the best of both worlds. However, for those wishing to remain entirely 'alternative', the school feels that completion of the course is adequate recognition of achievement. Comments by teachers are seen as more useful than grades, and the acquisition of knowledge and ability a desirable end result. A document of completion will be provided for each course, for those who require it, with notes on work done and general comments by teachers (a bit like work references), but a hierarchy of marking will not exist – in fact, this is one of the underlying principles of the school. The duration of courses will vary, and a teacher's (or teachers') expectations of pupils will be outlined by the individuals involved. Internal teaching credentials will be assessed by a panel of practitioners, scholars, and 'peers'. We are not interested in 'cheap and easy' degrees, but have a genuine commitment to learning, expressive creativity, and a non-invasive environmentalism. These are the end results we seek. We hope that a genuinely alternative way of educating might arise out of these general practices.

invitation

In essence, we are inviting you to participate in the School of Environmental Poetics and Creativity at its inception. It will be a long-term process to get this up and running, and it comes out of a vision – a necessity – on behalf of the founders. We don't have time any longer, and unless we are active now to halt the erosion of the world's environments (non-human and human), the end result is clear. So, if you think one of your pre-existing courses might benefit from tangential interaction (really, the offer to the students to participate at no expense to themselves), or a sense of mission eventually to teach and interact physically within one of our 'campuses' around the world, please let us know. There are also committee positions to be filled, and a lot of typing and grinding groundwork to be done. I doubt many will feel zeal enough for the latter, but it would be very welcome. The point is, this is about notions of personal and communal liberation, so nothing binds you other than a desire to participate – we are not attempting social engineering, but offering a genuine alternative. We wish to be innovative, forward-thinking and, above all else, respectful.

Addendum (2009): Disclaimers, Caveats, and Open-Endedness – A Rant!

I am not an ecopoet in the sense that Leonard M. Scigaj determines and distinguishes in his fascinating *Sustainable Poetry: Four American Ecopoets*. I think of ecopoetics, in his sense of the word, as *egopoetics*, and the witnessing of nature as 'a separate and equal other in dialogues meant to include the referential world' (11) as an act of convenience and not reality. As long as one engages with the materials of human 'progress', there is no way that a reality of 'equal other' exists. This notion of 'equality' might be 'sustainable' as an illusion (or delusion) of interactivity, but is about acts of responsibility from a position of human perceptions (or self-denials) of superiority. The privileged can never account for themselves as an equal other; they are simply creators of the other.

Neither do I 'reject the Derridean postponement of meaning along an endless chain of signifiers' that Scigaj argues is rejected by the four ecopoets he refers to (11), as that is closer to the truth of human mediation of nature (and of one another) than any claims to an authenticity, or a cyclical intactness of nature. Scigaj says, 'One might define ecopoetry as poetry that persistently stresses human cooperation with nature conceived as a dynamic, interrelated series of cyclical feedback systems' (37). Yes, I do think when it comes to ecological respect language is 'inadequate but necessary' (37), but I reject that human language is necessarily capable of being 'cooperative'. What is 'cooperative' exists outside of language (or purely inside the 'text', if one likes!). To respect nature doesn't mean I need to trust it. Language inevitably breaks down not only in the face of a damaged nature, but in terms of eternally damaged (and damaging) aesthetics.

Nature is a construct in language and its extension, aesthetics. Yet I deny aesthetics, and a 'school' of environmental poetics recognises *poetics* (in all fields and forms) as an act within the context of this position – that is, that poetry (for example) is merely postponement of meaning, and never meaning in itself. Even in its most activist forms – as declaration, outings, revelations, and agitprop – poetry is still mediated through an aesthetics (from shape to word-order: 'craft'), and thus a tool of placation (which is in the human need for contact with fellow humans, 'communication').

It could be argued that Ecopoetics (note the capitalisation) is an excuse for not making the big choices of abandoning those acts which violate and damage ecologies – a discourse that in many ways substitutes for direct action. We need to question the very use of devices and machines that allow us to participate in such discourses (we might think indirectly of that real environmental poet who went beyond a deep ecology, R.S. Thomas, in writing against the farm machine, but affirming nature and the validity of human cultures at once). Ecology is a science; ecopoetics is a faux science. (Note: 'Contemporary ecopoets', Scigaj claims, 'usually become interested in responding creatively to the broad outlines of ecological thinking; they seldom write poems that demand particular knowledge of the technical intricacies of scientific theory' (12). This is, in essence, a denial of many linguistically innovative poets' modus operandi, including the work of Cambridge-based J.H. Prynne, despite the mild qualification Scigaj makes after this claim.) Science is not of nature, it is anti-nature. Poetry as science is anti-nature, but it can be ecological. The use of the 'word itself' to stand or substitute as referent, but also to delay overall meaning while still contributing to 'impression' (ultimately, what humans in their mediated contacts with 'nature' rely on – a sensual reaction to what is perceived as the non-human 'created' from the smell of a flower to the sublime view from a mountain...), shows the triumph of language over content. Traditional 'nature poets' are often no more or less in contact with the realities of human damage and disturbance of the cycles of nature than are 'ecopoets'.

Ecopoetry is about how best to sustain human impressions of nature, and not to lose them as an intactness of human experience. Nature is seen as enriching – as a separate world to be respected, but one in which the human overlaps and has responsibility. I don't see it this way: I see no responsibility because the human is of nature and the moment s/he makes herself superior to this s/he no longer has a vantage point that's reliable (can it ever be otherwise?) – and nature is seen as deserving of intactness. There is a contradiction there. In the same way that 'national parks' should be left to themselves and unvisited (devoid of any human contact; or only human contact that shows traditional non-exploitative usage?), one asks whether there is such a thing as nature (on earth) outside the human. Antarctica: case in point?

The school seeks to explore the problematics of these issues. It is not a school of ecopoetics per se (doubting the integrity not of individual ecopoets but of a broader discourse) which is just the latest cop-out in the final gasps of domination of the land and the guilt it engenders. To quote in order to refute Scigaj yet again: 'human language is much more limited than the ecological processes of nature and uses of postmodern self-reflexivity to disrupt fashionable hermetic treatment of poetry as a self-contained linguistic construct whose ontological ground is language theory' (11). But this is a 'red herring' (another humano-centric adaptation) in so far as all poetry is language theory, no matter its ritual or inward-looking origins. Language comes out of 'ecological processes' because they are scientifically constructed views of nature in any given place – it becomes impossible to separate language (poetic or otherwise) from the ecological. 'Nature' as construct is surely not the same form of discussion point when you are 'inside' it: referents are realities and have real implications in terms of survival, and all genuine 'caring' about environments is conditioned by the mode of publishing and broadcasting one's views on *what* constitutes those 'ecological processes of nature'.

The school is a school of witness: we are inside the disaster of our own making. Discourse itself helps make the disaster. Investigating this paradox which is no answer in itself is the *raison d'être* of the school. Abandonment of all such forms of education, and of poetry itself (in terms of the printed page and the machinery that goes around the page: as opposed to orality, painting the body, marking the environment in a way that doesn't harm it...), is the suggested outcome: the pedagogy that is being offered through trial, error, and more error. Derrida offers more possibilities of truth in his endless chain of signifiers: the process is more honest and reflective of who 'we' might be in terms of the damage done, to 'nature' and each other. Scigaj claims, 'Ecopoets distil ecological process into aesthetic techniques to restore our lost sense of connectedness to the planet that bore and sustains us' (12). Should we, given his status as a definer of 'ecopoetics', accept his description of functionality...?

'Connectedness' is a fetishisation of belonging and exchange wherein 'country' has been relegated to the position of referent within a discourse on beauty and ugliness (the non-sustainable). It's like the Robert Bly-oriented 'expressive men's movement' and its false paradigm of affirmation of masculinity through various forms of connectedness. It is not surprising that Scigaj's four poets are all male (A.R. Ammons, Wendell Berry, W.S. Merwin, and Gary Snyder) – gender is the 'recessive gene' in much of eco-debating. It can't be: damage and gender are irrefutably linked and the school engages this as point of departure in discourse as much as any other factor.

The irony is that whatever poets want to believe, poetry *is* textual, and if it's to have an activist form, it comes out of the recognition of its limitations as much as its power to alter a textual landscape.

Coda: Visitors

In some ways I have been increasingly cutting myself off from contact with the 'outside' world. I still get out and about, travelling by train and (soon) ship, and turn up to lecture or teach classes, but these are discrete spaces that belong to their actions, and not places of social choice or growth for me. I am increasingly focussed on the block of stony ground on a hillside we have been making our home, in the region I have always considered my home place, though I have had many homes (especially in Ohio and Cambridgeshire), and the reference point for so much of what I write. But it is not my land, though the government (and as an anarchist, I neither believe in nor respect any government) might defend my right to 'ownership' against my beliefs and wishes.

The land I write is the land of the Ballardong Nyungar people, and I know it. At best, I am a custodian in a place that needed no other custodians spiritually, but in the grim light of rapacious development and exploitation, needs whatever protectors it can get. 'Our' place, Jam Tree Gully, as we call it, is consciously a Walden for me, and my writing there is now in dialogue with Thoreau. I have chosen *Walden* for a lot of reasons, not least because despite his limitations in appreciating the impacts arising from the theft of land that belonged to others, and the implications of the rhizomes of an earlier (and ongoing, of course) relationship to the land by Indigenous peoples, Thoreau knew what it was to step consciously outside society (though he was never in reality far from it). Also, because I have lived in America and had a son born there, in a country that gives citizenship by place of birth rather than heritage, it becomes a means of dialoguing with a place a long way away that is an essential part of my conceptualisation of the 'world'. The macro and micro operate in conversation as I inflect a colonised Australian space and my part therein, and a (very specific, it should be noted!) colonised American space and my meetings therein.

As I plant trees, climb the steep hill and watch the wedgetailed eagles, stand watching the kangaroos in the early morning or evening, I bring to mind the woods of central Ohio near Kenyon College, Mohican State Park, which I often visited with family and even on one occasion with my students, and the separations and links declare themselves loudly. The politics aren't global but local, and what I remember are the 'isolations' of each melded into a self-constructed commonality of abandonment – of places to avoid the social activity of the capitalist marketplace. Financially, socially, pragmatically, Jam Tree Gully doesn't make sense, but it makes ecological sense, and it's ecological sense I wish to make.

My poetry is inseparable from this process. When I was a teenager writing poetry on my uncle's Wheatlands farm, I believed that learning the names of things (in as many different ways of naming as I could discover – the common name, the name we kids made up, the Latin names, the local Indigenous names) would make them part of me and me part of the place. It was like an entrée to belonging. I collected proper nouns. I was wrong – that's not belonging, it's visiting. Names evolve with belonging – you acquire them without work, even without consciousness. They are pre-speech. In my poems, I deploy these names still, almost obsessively, though it's not to pin them down as an inventory, but rather as a way of acknowledging genuine strangeness, an alienation of my own condition from theirs.

The politics of *de*-anthropomorphics – by which I mean the poet fearing anthropomorphics and so avoiding descriptions of animals and plants, because to remove the discussion from a register of human comparison is ultimately impossible (we always get back to it via degrees of separation) – has led, ironically, to my mind, to a limitation in the ability to empathise with 'nature' outside the human. The pathetic fallacy is actually a generator of positive engagement with nature, however impaired and limited, and will inevitably be part of a communion with nature. I should note that this is a very conditional defence of anthropomorphics and the pathetic fallacy, largely arising from lament for other paths of engagement with the natural world being greatly ignored or unrealised. Humans often need to give human attributes to the actions of the non-human in order to transcribe into a language of affirmation and protest (to translate, as many other poets and critics have observed – but it's more than that), and when this is done with awareness of the politics of appropriation it becomes communion rather than yet another example of an egotistical sublime. The problem, I feel, in writing a poetry of place, is that few writers accept the status of visitor when it is inevitable that that is the role they are playing – from the overt visitation of the coloniser to the movement outside one's familiar space so that the familiar fades and changes, and the world becomes strange. This is the energy and drive of a poetry of questioning, as opposed to a poetry of ritual. The two aspects are part of a poem for me.

In *Walden*, Thoreau is highly conscious of his literal visitors (and their absence!), as am I. I have invited very few people out to 'our' place, though when I do, it's special and is about conversations on many planes. Recently, a friend who is a literary theorist visited there, and he was the one, more than ecologically minded activist friends, whom I took up on an offer to plant trees. His gesture said much about the dynamics of guest–host relationships, but more, that he was willing to step outside his comfort zone as he understood that having a visitor there at all was a major step outside my comfort zone. I wrote him a poem, which is part of this 'Walden' cycle. That poem is below, and really explains what I am getting at, but beforehand let's consider *Walden* and the politics (and spatiality, slightly) of visitations, which comes (interestingly and ironically) from the chapter entitled 'Former Inhabitants and Winter Visitors'.

Thoreau notes previous inhabitants of the woods as he understands habitation, and in doing so creates a map of both connection and disconnection. This is relevant to Jam Tree Gully insofar as the original inhabitants were driven from their land and it was claimed and divided up into properties by the early settlers in the context of the colonial government of the Swan River Colony, and ultimately the British Crown. The land 'our' block is on was part of a very large settler land grant with the landlord largely an absentee – this may be part of why it wasn't cleared as extensively as other areas (it is still quite wooded) – and Indigenous names and knowledge pertaining to it remain to a greater extent than for some other places in the region. I draw attention to these disturbing lines in *Walden*:

> or like that early settler's family in the town of Sutton, in this State, whose cottage was completely covered by the great snow of 1717 when he was absent, and an Indian found it only by the hole which the chimney's breath made in the drift, and so relieved the family. But no friendly Indian concerned himself about me; nor needed he, for the master of the house was at home....

Here, there is a tacit acknowledgement of an authority of presence prior to the 'presence' he tracks as relevant to his own, but gentle irony doesn't distort the impact of ultimately erasing Indigenous presence. However much this is imported into a broader discourse of acknowledgement and even respect, it remains the work of a coloniser, just as my work, regardless of its empathies and even the associations and culturality of family (early 'settlers'), must ultimately remain that of a coloniser as well. I should clarify here: I have not met my distant Kinsella relatives who I'm told are Nyungar, but I am acutely aware of them and desperately want to connect if the occasion should arise outside a 'contact' situation which would be patronising to all concerned, especially given my politics of total and utter belief in the primacy of land rights – that is, the return of significant tracts of land to Indigenous custodianship.

Thoreau's description of the dynamics of a small room, and how the act of visitation functions, is fascinating. But more to the point, for now, is this:

> The one who came from farthest to my lodge, through deepest snows and most dismal tempests, was a poet. A farmer, a hunter, a soldier, a reporter, even a philosopher, may be daunted; but nothing can deter a poet, for he is actuated by pure love. Who can predict his comings and goings?

In my case, it was the height of summer, and the theorist came in a car playing Aretha Franklin on his stereo (maybe along with some AC/DC?), and what I appreciated about his visit was, as I have already implied, that it was done within the knowledge that for me this was not usual – that I live as I do because I enjoy the company of my own thoughts, my family, and the ecology around me, within a framework of waiting and expectation in terms of how my custodianship (part of a custodianship) will be passed on, mediated, and reformed with the truth of its past – that is, its theft. His brilliant literary theorising (check out his *A Derrida Dictionary*) is, to me, a pure form of poetry. My friend, Niall, is supremely *untranscendental*, which makes his willingness to bring his family to plant trees all the more resonant and creates an ecology of friendship and visitation as much as an ecology of land. So, anyway, here's that Jam Tree Gully/'Walden' poem written for my visitor:

An Elective of Gradients

You choose which inclines you show a friend,
or which inclines your friend might favour –
but he makes his own way through the stones
and up the steepest parts and is interested
in what happens when water runs and cuts.
He is interested in gradients and erosions,
in the pair of eagles that come at dusk
before shutting down, in the echidnas
eating the termites that hollow York gums
that 28 parrots nest in. He is interested
in bringing his boys up here to plant trees,
to labour. I offer to pay them and him
but he declines, saying he would like them
to labour where the steepness sharpens
seeing and their work will grow without end.

APPENDIX 1

On Anarchism: Tracy Ryan Interviews John Kinsella

[This is an edited transcript of a discussion at the State Library of Western Australia, 2004, originally with introductions by Professor Bill Louden (formerly of Edith Cowan University) and anarchist activist Mar Bucknell. – JK]

Tracy: There are many different historical aspects of anarchism, some of which you would probably define yourself against. And I'm thinking particularly of the aspect of violence. So I'm going to start with a quote that I read recently in a postgraduate French course studying Sartre's play *Les Mains sales*, in which a lecturer gave the following definition for the benefit of students:

> An anarchist is a person (this is a very dated definition) who seeks to overturn by violent means all constituted forms and institutions of government and society, with no aim of establishing any other system of order in place of that destroyed.

Could you comment on this in regards to your own understanding of anarchism?

John: Today I don't want to get into an historical discussion of anarchism. I don't want to regurgitate nineteenth-century anarchists, but Malatesta [see his *Anarchy* (1891) – NL] made a great point about anarchism being the abolition of government and not the abolition of society, and I kind of concur with that. The replacement of governing institutions and hierarchical bodies of control with co-operatives, with mutual aid organisations, people interacting to support each other. It's a very viable and practical alternative to me, if not the only alternative. So from the start what I'm talking about is a world without government, and not a world without social institutions or interactions. 'Institutions' is the wrong word – social interactions. I think that's extremely important.

Obviously I would say that such a definition as you've just quoted is absurd. A dictionary definition that's very convenient, immediately isolates, and removes any debate about anarchist issues. Anarchism isn't, and from the most aggressive anarchists I've never heard it put in such terms, simply a violent overthrow of the State with nothing in its place; that's nihilism and not anarchism. I'm a pacifist above and beyond everything else, and veganism, the kind of non-use and non-abuse of animals, is the basis of my anarchist thought. I start from there and move out. So an anarchist's world is one in which animals are equal, if you like, as much as humans.

The very starting point of a violent overthrow is not possible from my point of view. I don't believe in 'revolution as such'; I believe in change by example. And I'm going to be referring, as I know Tracy will as well, to a guy named Colin Ward. This is a book of his that's just come out, written with David Goodway, called *Talking Anarchy* (2003). Colin Ward is an interesting British anarchist, who is very much involved in architectural solutions to housing for people. And his anarchism is a very pragmatic and a very practical anarchism that works within the context of the existing State. He believes that the State can be best changed by good example. So if you behave in a way that's better than the government is behaving, then people will gradually see that as a viable alternative to living, living communally. There is a lot I disagree with in Ward, but that basic principle I really do agree with. So, just as a starting point for this, I totally reject any kind of violent overthrow of anything – it seems a contradiction in terms to me.

We are working towards a better world of egalitarianism and equality especially in terms of what people have or don't have. Then the moment you introduce violence, you are introducing a hierarchy already. Violence is the ultimate form of hierarchy. It's the most controlling form of hierarchy. I would argue that many actual actions, ethically, cancel out any positive results. And I think that is a very personal view, as I said, that comes out of pacifism and veganism. That's where I start.

Tracy: What do you make of the common complaint that anarchism as an idea is too optimistic or utopian? Does it rely too much on the concept of human good will and altruism? Is it naïve, vis-à-vis the selfishness and violence we see enacted us daily, I mean in the world that 'is' rather than the 'better' world? A similar accusation is often made regarding pacifism.

John: Well, no, you see, I don't see anarchism as utopia. I see it as something incredibly realisable. Speaking of Malatesta [see for example his *Revolution in Practice* (1922) – NL] again, he also made the point – you can't expect us all to know what will happen after changes come. He said *after* revolution – as I've already indicated, it's a term I have a lot of problems with for all sorts of reasons, which might become evident as we go on. But certainly the point is that people and things find their own level of interaction. The basic principle of mutual aid, which is fundamental for understanding anarchist thought, is that people naturally help each other because it is in their best interest to help each other, and when Kropotkin [see his *Mutual Aid*, published in 1902 – NL] wrote about mutual aid, he looked at the false nature of the Darwinian model in which people compete to beat each other, which is the system capitalism is based on, and basically, eventually, some of the strongest will survive and conquer. The other way of looking at it, he says, is that animals have always had to work together to survive in their communities, and he gives interesting examples of that.

This kind of mutual aid is fundamental not only to the existence of animals, but also to the existence of humans. We don't have a policeman sitting watching over us – we might have in the audience! – we don't have someone policing in this room, and people are sitting here listening to an argument that some may not agree with. However, basic mutual aid is there in the sense of co-operating together to actually hear something, and maybe express ourselves at the end.... People can interact, people can respond in a responsible way towards each other without being told to do so, and the idea is that without governing bodies, without the judiciary, without the different constabulary, without the kind of legislature and these kinds of things, we can actually do this anyway. So there is a very practical side, certainly, to the anarchist thought that I'm interested in.

The other thing is that, apart from these 'what will happen' scenarios, the idea that we are fundamentally good and not bad drives my thought. Many people say that a lot of the stuff I write about seems very negative. Well, negative in the sense that I criticise things I see as wrong. I see the treatment, the locking up of refugees as wrong, I see the war in Iraq as wrong – I can list the whole series of wrongs I feel. My subjective take on it is one thing, but fundamentally what is being talked about here is an ethical way of living. Now, you may have very different ethical views, but at the same time you have some form of ethical system where you denote good and bad, and I think intrinsically people have that ability, and I don't think it's a class ability, and I don't think it's something you are taught. I think it's very inherent. I think that in the same way that an animal is neither good nor bad, they simply *are* – I think that humans are neither good nor bad. They are inherently good because of that – if you know what I mean. You can have both because both exist in you; goodness exists in you as well.

So my kind of take on it comes from an ethical standpoint, and Bakunin is a problem for me in a variety of ways. I strongly believe that many ills of the world have come about because of the abuses of science. I have no problem with the accumulation of knowledge and the use of knowledge; I do have a problem with the systematising of knowledge, when knowledge is used to create a hierarchy, and science more than anything else has created hierarchies. You know the addiction of discovery isn't always to benefit human kind; it's to benefit the 'discoverer' or the 'culture' of discovery – the actual pleasuring of discovery – to 'enlighten' becomes to fetishise ... in a marketed sense to add value to science. Now Bakunin recognised the irony of science but he still, like many anarchists in the nineteenth century and the early twentieth century, and indeed some people I would consider having very similar views who were involved with Freedom Press in Britain, strongly supported the kind of scientific approach to an anarchist solution.

So they say that when the government is gone, what we will have is a kind of knowledge; science will continue and the hospitals will be hospitals, and so on. And I have a more agrarian view of the world, and I think that society would naturally break down into smaller component parts. They are talking about a decentralised world, obviously, but I'm talking more literally about a very small component part, a very agrarian kind of anarchism.

I believe in direct contact with and the preservation of the land, and I think the only way you can do that is by actually understanding what the land physically is. Bakunin makes this very interesting point, and I strongly disagree with it. He says: 'What I preach then is, to a certain extent, the revolt of life against science, or rather against the government of science, not to destroy science – that would be high treason to humanity – but to remand it to its place so that it can never leave it again' [*God and the State* (1871) – NL]. On one level he is talking about the fact that governance isn't just literally the bodies that govern us; it's the way systems operate, and science becoming a system. The kind of move towards cloning a human being isn't about the survival of the human species, it's about a kind of value-adding to humanity, which is a very big problem for me. Bakunin is challenging that, as he would if he were standing here, I imagine, challenge the art of cloning and these sorts of things. But he is still reserving a place for science as rational. Science for him is rationality; it's an Enlightenment thing, and he believes very much in a kind of logic. Anarchism for Bakunin was a very logical thing. His logic extended to violent revolution – mine doesn't, and that's where we strongly part; but it's an interesting point.

Tracy: Science, rational science, as we see it practised, has a lot of problems of course for the vegan, in ways that are perhaps obvious; and also, thinking of mutual aid, there is a kind of implicit recognition there, even if it's not followed all the way through, of the relation between animals and humans, because Bakunin is arguing that it's not only in the competitive sense that we draw on our animal origin, but also in a sense of mutual aid.

Thinking about veganism: in my experience, many people, if they have heard of veganism and know what it is, believe it to be a dogmatic and authoritarian type of outlook. This may be partly a fear of difference, because it is potentially threatening to meet someone whose lifestyle implicitly rejects your own, and especially when it comes to food, because eating together is such a communal activity – some feel very rejected by someone who doesn't eat the same.

But it's also true that some vegans are very loudly critical of non-vegans and may appear to be rigid and fanatical. Could you comment on this image – how it squares with your own personal outlook, and also with your anarchist beliefs, because that kind of rigid authoritarian stance would be directly in opposition to anarchism.

JOHN: One of the things that has always disturbed me in talking to fellow anarchists is that as soon as you mention you are a vegan (most anarchists I've known haven't been vegans or vegetarians – some have, but most haven't) the objection is that you're imposing just another authority, another kind of hierarchy. But if you see equality between animals and humans as a starting point, it's not possible to have a hierarchy of imposition or declaration in that 'equation', and that's the way I see it.

One of the scariest moments of my life when I was at Cambridge, where I lived part of the year, was when I met my first vegan straight-edge. He was a young guy and he had engraved in his arm a statement – it wasn't 'meat is murder' but it was something very similar – he did have some line from The Smiths on his arm, and I got talking to him. He went through his list of absolute 'don'ts', including no sexual activity – basically it was no pleasure on any level, and intellectually I can appreciate where he was coming from, I could see there was a kind of abstaining that was almost the ultimate control, and I can respect that even if I don't subscribe to it. But where I don't agree is that he said basically: those of my friends who have betrayed us have been branded, physically with a branding iron, and I said to him –'That's not a vegan activity. Vegan activity is not to damage or utilise or abuse any animals, humans included, for your benefit or for any other purpose.' And he said, 'No, we don't see it that way. We see the only way of creating a revolution against the damaging and hurting of animals is to be really strict and rigid in our thinking.' I could see where he was coming from, but it immediately defeated the ethics of his veganism, for me.

Near Cambridge there is something called Huntingdon Life Sciences, which is the main place for animal vivisection and animal research, and straight-edgers have been very prominent in direct action against this, literally breaking into it and damaging it and so on. Now I can understand, and in my younger years I was involved in direct action against property, but I learned that this can develop its own hierarchies of behaviour and overwhelm itself with contradictions. I have felt over the years that violent direct action – even against property – delays but doesn't solve.... I was prepared to try anything to resist and challenge what I considered to be State-imposed oppressions. I thought, well, maybe it's worth trying. Mostly it was yelling and screaming at the tools of State and corporate capitalism, which I was very inclined to doing, certainly around the time I met my anarchist friend, Mar Bucknell. When I met with Mar and 'his' group in Fremantle, it was a really amazing experience. I had all this intellectual stuff, I had read all this stuff and I thought – this is where I'm at, this is what I think. But what struck me with these people and what deeply interested me in them – even though my behaviour was reprehensible and I apologise to them for my nihilism at the time – was their dedication to idea and action operating in tandem. Everything had to be discussed and worked through as a group of individuals. A consensus operated, as opposed to my individualised martyrdom for what I believed to be right.

This is twenty or twenty-one years ago, it would be 1983/84; and as Mar told me at the time I was a nihilist, not an anarchist. I just wanted to basically remove everything that was a problem very rapidly. What interested me about this group was that they actually had practical solutions. They had this wonderful newsletter (and this is how I discovered them, when someone gave me a newsletter, *News from Nowhere*) where I saw this thing about someone who was in academic life, who was an anarchist, who'd drawn up this plan for a possible agrarian commune – with practical outcomes like 'that's where the water tank should be' – the kind of background information needed, and it made sense and it was very practical and I loved that practical side to it. These guys were very practical – direct action to them was something that was mediated by a kind of a longer view of things and that impressed me. Although I didn't fit in, because I wasn't capable of fitting in with anyone, pretty well, at that stage, I have thought about it over the last ten years and tried to put a lot of those things into effect.

Tracy: I want to pick up on that idea you just mentioned – at that phase of your wanting everything to be sort of gone, destroyed, whatever. The anarchist writer Colin Ward, whose book you have already mentioned, suggests that the people who most readily attack the ideology of non-violence are those with little experience of the ugliness, squalor, and arbitrary nature of violence; so he's saying those who say that non-violence doesn't work are the ones who really haven't had much to do with violence. Could you comment on how your witnessing or experiencing of violence may have influenced your vegan, anarchist, pacifist beliefs? I'm thinking of poems of yours like 'Shootings' (see *The Silo*), in which you write about early experiences with animal death on farms, but also of any other experiences that you felt were crucial.

John: I was pretty aggressive. I had a major substance and alcohol problem for a long time and I was an aggressive person, and I tried to deal with things very directly, and very ineffectually, in the long run – I certainly learnt that. I suppose having a sense of aggression about you makes you think about it generally, but more than that, I grew up shooting everything. I had guns, and on the farm – when I spent time there – it was considered to be what you did; you literally went out and collected 'trophies', and that kind of trophyism was a very big part of my life up to age sixteen. I shot everything that walked, crawled and flew – that's what I was. I shot the things I liked the most, I loved 28 parrots, they were beautiful; I shot them because they were there, and what was disturbing about this was that I actually understood what I was doing. I wasn't some kid who was conditioned to this; I pursued it as an art form because it struck me that it was a kind of masculine thing that someone who was very directly non-masculine as a kid could do.

It struck me as a way of kind of identifying with that part of the culture I was invited into, so it was that kind of violence, and I stopped this because of two incidents – one when I saw an animal chew off its forepaw because it was entrapped and it chewed off its entire limb trying to escape; and the second was when I hit a ram driving past a ute on this gravel road near the farm. I was actually eighteen when that happened – I was farm minding at the time. I shouldn't have been driving, but that's another issue – I hit a ram that ran out and fell over a fence and ran just in front of the vehicle, and it had a broken neck, and I was terribly distraught, as I didn't know what to do with this animal, and I shot it in the head and killed it, and pretty well everything changed for me from that period on. It took a while before I became a vegan, but that event really stood out.

Also as someone involved in that kind of lifestyle of drugs and alcohol, I saw a lot of violence, a lot of serious violence, and it used to repulse me. So my activity against violence, my pacifism has come from experience rather than from just a concept. I have experienced a lot of violence as an addict, but also since cleaning up my act – as a pacifist being physically challenged for being a pacifist. Being tested. People are so affronted by non-violence, it's even a far more effective device than violence anyway, and if you want to bring change, you can bring it most effectively by not biting back. So my non-violence is something that's come through a kind of fire if you like, and there are many other things I've seen over the years that confirm that, especially in other countries – that I won't go into, but I've been involved in the middle of things on a number of occasions where people have been literally fighting with guns and stuff. I found myself in the middle of battle in the mid-'80s and saw people shot. It changes you.

Tracy: Your veganism connects you in other ways to the environment. Is an anarchist necessarily an environmentalist, and what forms of action then, or attitude, might that take?

John: I've not met any anarchist who is not an environmentalist, but as Colin Ward points out, they are most often environmentalists of the urban. Certainly in the London anarchist groups, as you would expect, in the British anarchist groups, they are very urban-centred, big populations in urban places, and Colin Ward talks about the urban environment as someone who is concerned about how people are housed; it's a very important thing to him. I have an interest in that, but I am very much what you would call an ecological environmentalist.

We've got the anarchist activists in the tuart forest in the south-west, and we need them, and that's good. But my argument to them, when I was talking to them – people I feel deeply akin to who are pacifists as well, but they were doing a lot of locking-on to machinery and I said that's a very pacifist resistance, but it's a very finite effectiveness once you're locked up. You've been locked up a few times, you are disempowered; the State will get you, believe you me, it does. And then the defence is gone, then the trees get knocked over, and as I was talking a tree was going down. I think the solution to a lot of these problems is one in which we use our mutual aid and work together. We live in a world that is not anarchist, unfortunately, and we must work with people who aren't anarchists and may have very different views, and I don't have a problem with that. I'm very pragmatic in that way: I'm quite happy to work with people for a cause and work together and have a number of different approaches to a situation, because I think if we get very monolithic in the way we view a problem, then we are really serving our own interests, and our own emotions, rather than the actual cause of, say, saving the forest. I'm not suggesting these people are serving anything but the cause of the forest; I'm saying there has to be a more interactive approach to the preservation of environments because you can save something for six months, but then six months later....

We went up to the Avon Valley National Park, not far from Perth, a beautiful park where the Avon River runs through. You only get to see it basically when the Avon Descent boat race is on and people go and watch them going down the rapids, but Boral, the large mining conglomerate that mines stone – their mining operations are based on the edge of the park, possibly going into the park, and that's an issue, as well, that I'd like to find out about. The government doesn't survey its own wrong; it surveys your wrong according to itself, and that's a truism. It strikes me that one of the most effective pacifist ways of dealing with governments is to kind of legally dismantle them from within, as with defending the tuart forest: get a few really good lawyers down there and get those boundary lines checked out where they have violated the lease and get them, and gradually the process of decomposition takes place. So I'm into very practical solutions.

Tracy: Malatesta says that we must make ourselves 'able to live anarchistically. The remedy does not lie in the abolition of organisation but in the growing consciousness of each individual' (*Majorities and Minorities*). Is it really feasible? You've just been talking about ways in which we might live as anarchistically as possible under that umbrella, even when we differ from it. Is it really feasible; are we inevitably going to get drawn into complicity with the State?

John: I actually have such a negative view of the State, that it's so oppressive I don't think we could ever be complicit with it, because it's always going to get us in some way or another. The welfare state is obviously set up in a paternalistic way to protect people, theoretically, but the further you move away from who is actually doing the governing, the process of representation, the less you are going to be represented. Democracy for me is not a free society; it's the opposite because you're abdicating your responsibility and your right to have a say in how you live. My local government member for the York region – he doesn't know me and I don't know him – we might meet socially, but the point is I've abdicated my responsibility to him.

Democracy is not about giving you your rights and freedom. It's about working through a totally delayed and distracted and deferred system of response to needs, where responsibility is in essence entirely removed. The more 'welfare state' it is, the more paternalistic it becomes, of course. On the other hand, thinking about Ward.... Thatcherite individualism, for example, yields right-wing selfishness. Ward notes certain things that were in some senses more anarchistic than the Labour government that followed Thatcherism, and that's not to say he supported Thatcher – he loathed Thatcher and campaigned outspokenly against her, but the idea, for example ... the example he gives is when council houses were sold at very cheap rates to the occupants because basically the Thatcher government did not want to spend any money on the upgrade and the upkeep of these houses, so the houses were sold for £30,000 – half their price – to the occupants. So suddenly a group of relatively underprivileged people had property, and unwittingly what the government did was empower 'working class people' through property ownership which they normally wouldn't achieve under the rigours of wealth marginalisation, for want of a better expression. Ward picks up on the point that governments actually don't understand what they are doing when they are working in other directions and the Thatcher point is a very good one, there, in every way.

As regards individualism in America, of course, historically, there is a whole thread of right-wing anarchism, individualist and liberal anarchism that is very much packed into the 'what's good for me is best, and bugger everyone else' school of thought. But it's a lot more complex than this. The American dream of doing pretty well what you want without government strictures, at least from the 'Feds', or without interference, is so much tied up with property – a kind of liberty through the rights of property accumulation.

Umbrella anarchism is a co-existence with but not an approval of the State. Anarchism on a 'micro-level' can bring change in quite dramatic ways. An umbrella both protects and deflects (literally, from rain). The anarchist is protected from the physical abuses of the State's legal and military apparatus by 'co-existence' on the least directly compromising levels (purchasing food, use of water and hospitals, and so on), but is also deflecting its intrusions by making use of facilities and means outside the State's control (and corporate-State capitalist control) as much as possible (not banking, exchanging and bartering where possible, growing one's own food, capturing one's own water, refusing to vote, being involved in public and private protest, deschooling – I am thinking of Ivan Illich here – and so on).

TRACY: Malatesta again:

> in the contest of centuries between liberty and authority, or, in other words, between social equality and social castes, the question at issue has not really been the relations between society and the individual, or the increase of individual independence at the cost of social control, or vice versa. Rather it has had to do with preventing any one individual from oppressing the others; with giving to everyone the same rights and the same means of action. It has had to do with substituting the initiative of all, which must naturally result in the advantage of all, for the initiative of the few, which necessarily results in the suppression of all the others. (*Anarchy*)

I think a lot of people who aren't anarchists have problems with the 'individual' and how it's balanced against society. Do you want to comment a little more on that, on your notion of what individualism is, for you?

JOHN: You have two functional notions of the individual: we have the individual that we know, where what we think and what we feel are very much connected to this physical body we have, and that's a very observable state. There is also the individualism that comes in how we interact with other people, within a social situation – say five or six of us might get together and watch television, and we have very individual views regarding what we should watch – regarding the matter of 'choice' ... all arguing what is good for us is good for others because ... and so on. Struggling towards consensus, creating a comparatively egalitarian and acceptable pattern of watching. So there are two issues there – one of consensus where we have agreed to actually do something together, and one of actually having an individual view within that consensus (and a desire to have our own way).

In a non-centralised world, it seems essential that property is held in common, and pretty well most anarchists would agree on that – there is not individual ownership to the point where people can have actually more than someone else, because we are sharing in a distribution of wealth. But it is a kind of wrong thinking for me, an illogical thinking, in that property as such, as a definition, shouldn't exist at all, as far as I'm concerned. The problem is in the emotion of ownership. The desire to fetishise the object and exclude others having a right to it. To invest it with a personal spirit that makes it exclusive. The existence of property would be contingent on rights of access: to have access to the things that are required or needed at a particular time; so I might have this book, but you have access to it if you need it, only if it becomes 'relevant' on a basis of need. I might have made this book, I might have sat down and made my own paper, written it up and have it in my own possession, but it is something anyone can have. Same applies to a shovel or bedding or any other 'possession'. Some 'property' would need to be used constantly, and this is factored in: clothing, eating utensils, health items....

It's a kind of public moment; you are not hiding it away and keeping it for your personal edification alone – there are different levels of possession and materiality depending on need. I think that there are very literal and very obvious and pragmatic solutions to these issues of ownership or not. Things are not only in common – obviously they are in common – but I think we have *access*. So if someone's got a rake, then I will use it and then so on. It is a very feasible thing. For me it's not a matter of what you own; it's a matter of what you share.

Tracy: I'm thinking in terms of how that's organised in a wider context, with your ideas about international regionalism, which you've written about in relation to poetry and landscape. This is the sort of wide picture rather than person-to-person view, because the world, so we say, is very 'global' now.

John: International regionalism, in a nutshell, is basically interacting with communities outside one's own, respecting others' regional integrity, and confirming your own identity. So it might be applied to a social group; it might be a geographical region. Obviously the integrity of tribal or nomadic social groups that have a differing 'definition' of region, that cross lines of other community identities, is respected. Respecting that, and at the same time opening communication between those groups where communication might be desirable, or allowing communications. Or silences.

So in this regional philosophy the possibility of lines of communication (visual, verbal, exchange – a variety of modes of interaction) is key. I developed this idea in dealing with poetry, in writing a very physical, a very regional poetry – I write (notice I don't say, 'write about') a very specific area – and being very involved in international discussion on how things might change or what we might do. So this kind of theory evolved out of a necessity.

Tracy: Within this philosophy, people sometimes ask – what about identity groups, Indigenous identity groups or migrant identity groups, and so on?

John: Well, this isn't an issue of identity-as-hierarchy, but identity-as-choice *against* State hegemony. 'Identity' is completely respected and surely it is logical that ethnicity and social groupings or beliefs connected to land are enough to generate a kind of social structure, rather than having a government tell you what to do? And that is a very interesting differentiation. Bakunin thought that everything problematic began with the concept of God, that we immediately start with the hierarchy. The point is being made that there is a hierarchy for relationship in the way we worship. Now, I am not suggesting that people should suddenly not worship, but what it means is that the power structure dictates how one worships or how one believes or how one has faith. I'm not serving anyone, they are serving themselves – it's a matter of allowing people to recognise that the 'church', for example, whatever religion we are talking about, is dictating to you how you will believe.

You are quite capable of discovering how to believe yourself. Through your experience socially and otherwise, there are a lot of other directions available to you. I think it's a good point.

APPENDIX 2

Dialogue on Vegan Ethics: John Kinsella & Tracy Ryan

OHIO, USA, MAY 2001

JOHN: I became a vegan almost 16 years ago. I'd been a vegetarian for a short while before that but the logic of the action caught up with me pretty fast. Once flesh went, it seemed obvious to get rid of milk and eggs – I was actually living on a dairy farm at the time with my brother and then partner. My brother said that given the torture that was all around us, it seemed hypocritical to utilise animal products in any way. We all became vegan that night. I had spent some time in India before that, and had been impressed by Jainism. But this was a tenuous link. A consciousness of the exploitative nature of living as an omnivore had long crossed my mind. I knew it but didn't want to face it. As with many issues at the stage of my life, I hid behind substance abuse. Six years ago I gave up all intoxicating and toxifying substances. For years while eating no animal products, and wearing no animal products, I had smoked tobacco and drunk alcohol that I know today was made utilising animal by-products. I gave up tailor-made cigarettes long before 'rolly' tobacco in an effort to overcome the problems, and never drank alcohol that I knew had animal products in it – beer with isinglass, Guinness with gelatin, and so on. But I knew if I looked deeper I'd find animal testing and other by-products. A bit like knowing some sugars are purified using bone charcoal. So giving up was a coming to grips with my own issues of addictions and facing up to compartmentalised hypocrisies. Some vegans avoid issues by avoiding self-investigation. We do the best that we can, but are always aware something is there – be it in car tyres, or in the carpet we walk on. But we must do our best to overcome these issues, and not give in to them. We go as far as we realistically can. Realistically shouldn't be a word just for making it easy. I am genuine about this.

TRACY: Yes, it's a spectrum along which we move, and I'm constantly discovering things I didn't know before, even after more than six years of veganism. For me (and I think for you too) it isn't a matter of striving to 'find one more thing' to avoid, but simply adjusting my behaviour and habits as I discover things. An example would be the one you mention regarding sugar. It didn't strongly become an issue for me until we came to the USA, where I could not take for granted that the sugar was vegan. Now that will affect my processes of checking, wherever I happen to be. I've read things on vegan webpages where people say it's too extreme to check every little thing – if we went that far we would have to stop eating veggies, since insects die when the crops are harvested. But that seems to me to be missing the point. I can't avoid eating altogether: I can avoid certain other things. It just seems to me that for vegans, that's a non-argument. Similar to when defensive meat eaters say, 'But what if vegetables feel pain?' If veggies do feel pain, then meat eaters are causing it to way more of them than vegans, since they eat the creatures who also eat plants.... It seems to me that most of these non-arguments arise because people want to distract themselves from what veganism is actually saying. I sense this because I was like that too – there must be ways in which I still am.

I can remember as a teenager, when I was a meat eater, thinking a veggie friend of my mother's was so extreme – 'doesn't she know meat eating is natural to us' etc. – because I accepted that what I was brought up to do was the natural thing! Also, I had read a lot of 'science' on the subject and it all seemed to say that we needed meat for protein (it always ignored the other protein sources, as if they were trivial) and anyway there was B12, and that couldn't be gotten around. So I thought.

Now, of course, I know that, like most 'knowledge', these supposed facts are produced from certain beliefs – science proves what you want it to, a lot of the time, and this is particularly true of dietary studies: people find what they are looking for. It's much more generally accepted now, though, that, at the very least, frequent eating of meat is bad for you.

Notice how I tried to avoid the issues by words like 'natural' and by focussing on dietary matters as an excuse. I was raised to think cruelty to animals was wrong, but not to see eating them as part of this cruelty. My family certainly would have said that factory farming conditions were cruel, and tried to buy more 'ethically' within their own system of understanding – e.g. free-range eggs. But we were not inclined to question the eating of the eggs in the first place, or the idea that the chickens or other animals were 'for' us in this way.

It was not until I had surgery at the age of 25 and suddenly saw my own flesh as possible 'meat', cut and stitched and right in front of me, that I even thought about being vegetarian. And for the next five or six years I wavered from that position (coinciding with a green or 'environmentalist' obsession; I became an avid recycler, afraid of what was happening to the earth) to one of 'semi-veg' as I then saw it, what I would now see on my own part as hypocritical avoidance in order to please others, because it was socially difficult to refuse what they ate. I think all this was still too emotive and psychologically driven, rather than rationally decided upon, to work consistently, though I can see too that it's better to do some of these things than none of them. I see vegetarianism as a step in the right direction, that's for sure, and even lessening meat consumption means that much less damage is done. But I would no longer see these as valid positions for myself.

John: By veganism I mean the total non-use of animal products and by-products. For me veganism is an act of non-violence. And all animals are equal. The suffering of a bee is every bit as significant as the suffering of an elephant. And yes, one death is equal to another. Which is not to say that I wouldn't make a choice that favoured one over the other, but I would recognise the consequences of my actions – that my conscience would be aware of the consequences. Those who use honey and call themselves vegans are hypocrites: it is an exploitation of another creature's product. In the same way that cow's milk is for cows, so are honey and royal jelly for bees. And, for that matter, beeswax. Veganism is not wearing the leather jacket from pre-vegan days, or having a silk shirt inherited from a best friend. Many dietary 'vegans' call themselves vegans even though they advertise such hypocrisies. It is only a word, but like 'om' it has sacred and specific connotations. It should not be abused.

Tracy: My main problem with the diluting of the word is that it muddies the concept for those who might otherwise go further. If people want only to go that far, that's their choice and fine, but why confuse, why not use a more accurately descriptive word? Surely a person who avoids animal products but uses some by-products (e.g. honey) is a vegetarian, not a vegan. That is not a moral criticism on my part of their choice to be a vegetarian! It is a wish to be clear and specific about my own further choice. I don't feel comfortable about using the word 'hypocrisy', except of myself, because I can't know other people's intentions. I would prefer 'logical inconsistency' or something like that.

When I became a vegan I thought long and hard about people's ideas of keeping the old leather and wool stuff until it had been worn out. Some said it was to 'honour the creature that had been so used', and not to just ditch the stuff – this seemed not to make sense to me. Later I thought it would be more honoured in, say, having its remains or parts laid to rest – bury the stuff! The other argument was a kind of environmentalist one – it was wrong and extravagant to waste goods already manufactured. So at the time I passed them on, donated to second hand stores, etc. I don't think I would do that now, but at the time it seemed the best thing to do. I am simply not capable any more of *not* seeing leather as skin, or, say, the feathers in a pillow as the bird's painfully stolen former covering. I could no more wear the leather than wear a dead person or other creature.... When I was a supposed vegetarian, I wore leather shoes, simply because I thought only in terms of what I ate; as I said, it was a kind of psychological *reaction* to the idea of flesh. I could link it to ethical causes, but they were not the prime reason.

John: Though I deplore all forms of animal abuse and exploitation, and the destruction of natural environments, I do not support any form of violence in resisting this. What I condone is passive resistance and what I call 'linguistic disobedience'. I believe in the power of the word to bring change. And to let someone strike you down will bring much greater change than striking them. Furthermore, I believe in cultural respect, and find it problematic dictating the terms of the relationship different peoples should have with the space they live in. But what we can recognise is the universal wrong of torture, cruelty, exploitation, suffering....

Tracy: I agree – and the inconsistency in using violence to fight violence is quickly noted by meat eaters, not only by other vegans! It doesn't make any sense at all. Lots of ethical/political movements are troubled by this non-issue (I'm thinking for some reason of the 'revenge novel' that purported to entertain feminists some years back, like, say, Zahavi's *Dirty Weekend*: there is a moment of exhilarating charge when an oppressed reader recognises vindication in the story – but at what cost, finally? As with gender issues, so with animal cruelty issues). It makes no sense to bomb or attack the 'wrongdoers'. It actually compounds the wrong, increases the level of violence, argues that violence under some circumstances is justifiable – in which case, we have no case against violence, for who is to decide the circumstances?

John: Regarding the argument that insects die in the natural processes of growing vegan food, my answer is a simple one. If animals die as part of their cycle of life, if there is no intention on the part of the human and all possible safeguards are taken, then culpability is not there. I do not eat animals or use them because I can live without doing so. It is a rational decision. I cannot live without using plants, so I (or someone) must sow and harvest them. In what I grow and what I buy, I do my best to avoid abuse of the environment or the rights of any creature – insect, bird, mammal, etc. Intention is everything. Our commitment is to increase our knowledge, to develop practices that reduce the chances of exploitation and damage. If, in our conscience, we feel that we are striving to improve awareness and to operate within the integrity of that awareness, then we are acting responsibly, honestly, and with decreasing hypocrisy. So it's rational, but also spiritual. It's a way of life. Control and power over our own decisions, but not over the environment or those – animal or human – living in it. I am an anarchist, a vegan pacifist anarchist – I believe that decentralised communal living with a combination of individual integrity and communal responsibility, looking out for those next to you who might be in need, to assist without the desire of profit, is most compatible with a vegan life. I avoid the word 'lifestyle', as it suggests complacency and comfort. The vestiges of the non-human world are too much under assault to allow for rest. Vigilance is essential.

Tracy: Yes, but you make it sound much harder than it is! True, there's an intense mental vigilance that comes with being a vegan, but even that gradually becomes a habit of mind, wouldn't you say, rather than a strain.... I remember this same pattern from when I first started the recycling thing: it was harder then, there was no community/council collection of recyclables, you had to travel a long way to take them to a depot – but once the concept had been incorporated into the life (I don't have a problem with the word 'lifestyle') it was not difficult. It made people laugh a bit, or think you silly to go to all that effort – but so what? So few people speak of the deep pleasure in those commitments, once they become part of your being – and not in a holier-than-thou sense. I love my food, I love experimenting with cooking, I especially *love* making meals for other people the best that I possibly can, to bring them pleasure, and specifically as a vegan the pleasure and surprise that come to them from realising that this delicious food they are eating had nothing to do with cruelty. That you can 'eat your cake and have it too', that veganism is about love and goodness and the bounty of the earth – if that doesn't sound too gushily romantic. I get pleasure from food as a vegan that I never got, that it never even crossed my mind to get, when I was a meat eater. Meat was like a drug that prevented me from seeing or feeling anything much.

I used to work, as you know, in a department selling cookery books, and get frustrated by the gourmet eroticising of food in many of these books, vis-à-vis a mostly starving world. I don't think, however, that it's wrong to take pleasure in those things which cause no harm and deprive nobody, and that's one of the other things about a vegan diet, it does not contribute to world hunger in the way that animal-raising for meat eating does.

Frankly, I just can't think of a single good reason not to be a vegan. People raise the issue of convenience, but whoever wants to order their lives around that one, anyway? We all arrange our lives as conveniently as we can manage – vegans included – but it's not the be-all and end-all of anyone's choices. It's only because we are not in the majority that things appear to be difficult for us. Veganism isn't difficult in itself: if there weren't meat, eggs and dairy industries, we wouldn't have to read food labels.... It's merely the result of going 'against the grain'. Veganism for me is exciting and fun, too. And colourful.

John: Being against all forms of violence doesn't mean one doesn't have anger. If I find myself yelling, I regret it and will, at least in recent years, walk away from an argument that's overheated. A vegan has the right to feel anger as much as anyone else, but verbal abuse is a form of violence. Obviously it is not in the same realm as physical abuse, but it should be recognised for what it is – aggression. To feel angry about the injustices of the world is a natural enough emotional response, and it can have positive outcomes if that anger is directed into non-aggressive responses, but most often it just results in damage to oneself, and limits the possibility of effective action. There are no circumstances in which violence is excusable, unless it is in self-defence, or the defence of others in a situation where there is no room for an alternative outcome. This response is hijacked to endorse the violence against people who are responsible for the exploitation and abuse of animals. But the premeditation of the act invalidates the 'defence' argument. The defence response I refer to is of the moment, when there is *no* other possible course of action. Campaigns to end animal experimentation and exploitation that come out of a premeditated plan of violence perpetuate violence themselves. Most often the people involved are using the 'cause' to express their own anger and sense of powerlessness. Power is in the self, and is shared between those who do not wish to oppress others. It's self-respect and self-confidence; it is a sense of obligation to others. It is non-predatory and non-proselytising. One changes by example, not force. Force builds resentment and retribution. If there is not a 'spiritual' aspect to a belief system it ends in totalitarianism.

Tracy: When am I to know that there is 'no room for an alternative outcome'? In practice I know that if someone attacked my child I would probably do anything I could to save her – but she is in the unique position of being the person for whom I have complete responsibility. Now we have a responsibility towards others, but it's not quite the same.

John: An engagement with technology can often seem antithetical to vegan living. It goes without saying that the use of electrical equipment is contributing to the destruction of the environment, is detrimental to all life. Most technologies do this on varying levels – from clothes production through to the manufacture of paper, to the home computer. This is obvious. Any form of mass production is going to produce waste that is not readily absorbable by the environment in which it is created. The movement of materials from their original location to another upsets ecologies, and the need for energy saps the planet. One could go on. I have lived without electricity, grown my own vegetables, and survived in the most non-technological way possible within the framework of a so-called Western democracy. Various intrusions by the government – rights to land, spray programmes, and so on, constantly compromise the condition. Increasing population, hunger for resources, create yet more pressure. And then there's the social factor of the neighbouring farmer's son wanting to shoot you for looking different, only to be prevented (if slightly encouraged as well) by the local police. You need them but you don't want them. Does the process of balance become compromise? By using technology to create community and decentralised interaction, to work together for the well-being of other species, outweighs the loss by using the technology in the first place? I tell myself this is so, though I know to abandon it altogether is the only option I will be left with. I engage with certain technologies because I believe they help disseminate information, to bring knowledge. That knowledge and artistic endeavour are far more likely to create a receptive atmosphere for change. I utilise it so that we might abandon it. How far can this equation be pushed?

Tracy: I really don't think 'technology' per se is bad or good; it's *what kind, how* it's used, and for *what ends*, surely, that decide its status. Whatever kind of human society you have is going to have some kind of technology – and I don't know that it's necessarily the mass nature of production that is destructive, although it usually is in our current world. Isn't it more the drive for profit and the inability to put other values first? Some 'energy loss' is always going to happen – it's just that it's happening in crazy proportions nowadays. We ought to be minimising it. Also I think that living as a vegan in the world we currently have involves accepting that in some cases there will be no cruelty-free alternatives, but using the alternative in *every* instance where there is one. And reassessing, too, what is seen as essential – e.g. were there no cruelty-free cosmetics, I would not see them as important enough to use. Where the world relies on computers, I mean really relies, and there are no totally cruelty-free computers, there may be options of non-use for some people but not for others. I don't see balance as compromise. I think we have to have just the right amount of utopian optimism to move us onward, and just the right amount of pragmatism to keep us alive in the world that is. It will differ from person to person.

John: Medicine and medical matters are always problematic for a vegan. Some obvious points to begin with: homeopathic remedies (of non-animal origin) are always sought, and diet is pivotal. However, there are some medical situations where the person loses control through external force, loses the ability to choose treatment (not conscious, delirious, etc.), or where there is no other option (other than suffering and/or death), than to be treated with substances that have either been tested on animals, are animal by-products, or directly manufactured from animals. I have taken medicines in full knowledge of their origins, but only under extreme circumstances. I do not defend such a choice and realise its moral implications. My argument (or the argument others have made for me) is that I can do more animals more good by staying alive. Which is, of course, avoiding the issue. I would have no hesitation in having my vegan daughter treated with such medicines if there were no other choice available and her health was at risk. And yet I cannot condone on any level the manufacture of such medicine. The Dr Hadwen Trust is a great example of an organisation working to overcome such problems. There are most often alternatives, if people want to find them. Where I do stop is with the use of animal or human body parts to sustain a life. Once again, I can't say that I wouldn't make the choice for someone else, but for myself I would refuse an organ transplant or the use of a transgenic body part. There is a point beyond which life is not worth sustaining. The offensiveness of such actions, the devaluing of the integrity of another complete body doesn't rationalise it for me. I do not wish to be made up of component body parts. This profit-making industry is singularly the most offensive of all ethical questions. That committees sit and discuss the gradations of what can or can't be done belittles the lives of the animals that are sacrificed to make it possible. And I'm not even talking about the vanity industry, which is another story.

Tracy: Yes: I wonder how many people who have collagen implants to thicken their lips for 'beauty's' sake think about the animal the collagen came from – or, in the US, sometimes the human tissue (via organ donations – unbeknown to the donor!) it came from? There's something vile and 'sacrilegious' in the broadest sense about this. I'm totally with you on transgenic animal parts; it's inexcusable.

I can't comment on one person's decision to donate an organ to another person, as that seems to me to involve their rights to do as they please with their own bodies. I can't imagine wanting, myself, to take someone else's organ, but it's not a position I've ever been in. I do think that once anything like this becomes an industry for profit it is just plain wrong. One should recognise that 'the integrity of another complete body' has connotations of religious belief that not everybody shares (body as temple, etc.) but then, too, the State ought not to override the rights of those who believe against them. As you know, recently in Western Australia there have been plans to make organ donation at death *automatic* unless you sign a form to say otherwise. I do think this is a major infringement of individual rights (the body may be dead but it is still no one else's automatic property) in the name of utility.

John: Such violations appear 'natural' because they belong to a history of violations. I do not believe in keeping pets, for example, though I do sanction rescuing animals and providing them with homes. To fetishise an animal, to turn it into an accessory, no matter the good intentions, is to deprive it of its rights. Even the most affectionate family pet environment creates a human-centred power relationship. I would never feed a dog the flesh of other animals – there are alternatives now, to meet dietary requirements, if need be. However, the situation should not have been created in the first place and should not be perpetuated. I cannot keep dogs because I cannot feed them what they need. The wild dog hunts for itself, and that's as it is. Obviously such conditions cannot be created for the domestic dog and so, as with farmed animals, they should enjoy their lives and not be bred to further this tyranny. I do not condone sterilisation or castration – it is brutality – but I do see that efforts should be made to prevent breeding. This in itself is a tyranny, but less of a one. These are things we have to think about as vegans. The family 'property' is turning into a rescue centre, and this is as it should be, but such acts have implications – we must bear these in mind, and not let our needs and manipulations be the centre of the activity.

Tracy: I don't 'disbelieve' in keeping 'pets', though I don't like the word 'pets', and I think the only justifiable form it can take is with rescued animals, animals already in existence – the breeding of animals for the purpose of their being our domestic companions is wrong, to my mind – it removes their freedom. I think that human-centred power problem you are talking about is all summed up in that word 'pet', the accessorising mentality. I knew someone who wanted to have four companion animals 'put down' because the human was moving house and the animals had become an inconvenience. I wasn't a vegan then but I was still shocked – I think I had failed to understand what the term 'pet' really meant. I do believe that in many cases looking after the companion animals may be the morally right thing to do, but not as a self-perpetuating enterprise – there would have to be a shift towards ending the breeding of animals for this purpose. Even many vegans don't see a problem with sterilisation (though it *is* a forcible cutting of the body – think of when it was done to some groups of humans without their consent, and we felt the outrage). Usually vegans who advise this are trying desperately to control environmental problems and problems of suffering in the particular large animal population. With some creatures, ones that would suffer greatly if not allowed to roam, these people might argue that sterilisation was a kinder 'violence' than curtailment of physical freedom, for example. It's a very, very difficult topic on which there are hot and opposed feelings.

John: If the spirit of veganism is non-violence and respect for all living things, then a refusal to accept the overt and tacit control of the patriarchy must be part of it. Not only gender equality, but also a recognition that both male and female are conditioned to work within the structures of masculinities. The hierarchy, gene control, the competitive Darwinian drive, the striving for spiritual affirmation by using the planet as a springboard, the control of the patriarchal family unit (and matriarchies are simply female replacement scenarios within a patriarchal structure) must all be confronted and challenged.

Tracy: 'Patriarchy' is a word I use as a kind of shorthand, but we have to (before we say anything more) accept that it's a word a lot of feminists now reject because of its blanket nature, the way it was used in cultural and radical feminism in earlier decades, possibly, to imply that all women's experiences were the same (and in the gender studies sense, the word *does* stem from this context). Yet it's hard not to have a short-form word to delineate the background we're talking about.

I think what Carol Adams and some other animal-rights-feminists say about the oppression of women and the oppression of animals as having common features (they go further than that, though) is very interesting. Certainly I think that feminism has a responsibility to address animal rights and if it doesn't, it has missed a whole zone of culture intimately connected to its concerns.

The risk we take in applying feminist thought to animal rights questions is perpetuating that old woman = nature association that people have – but they're going to have that anyway, and there are ways and means of addressing the issues without falling into some kind of 'earth mother preaching' trap.

John: The compatibility of different religious beliefs and veganism has always interested me. I always found it difficult to accept the animal cruelty of the Bible, and the metaphoric – or literal – eating of the body and blood of Christ. Though there is much I admire in Christian belief, and though I have strong respect for Christ, these factors always pull me up short. Of course, it doesn't affect my ability to believe, any more than seeing a Buddhist eat meat should turn me against the wonders of Buddhism, but it does make me sceptical of the Church. I do not trust any power centres, and the Church is just one of them. The Christian feminist theologians who are unafraid of the Councils, who utilise the gnostic texts and the apocrypha, who un-read the patriarchy, impress me far more. For me the Bible is not a static volume, but an ongoing, organic, and growing collection of writings through which the word of God is culturally interpreted, represented, and discussed. If all writings are inspired by the Godhead, they are also products of the experience of the writer/s. They are symbolic texts. A vegan can find solace in the Bible, in the Koran, in Buddhism, and the Bhagavad Gita, in Zoroastrianism, and so on. Veganism is both logical and spiritual.

Tracy: I would qualify that by saying that it isn't spiritual for everyone – some vegans would not identify with that potential element. I was raised in Christianity so I'm familiar, as you know, with the motifs you are talking about – they don't hold any problems for me, not the Communion, anyway. Reading of the sacrifice of animals in the Bible makes me uncomfortable, and there is much reference, symbolic and literal, to fishing and other forms of killing as well. However, the Bible is a mixed bag. There's a very strong anti-cruelty message in there too. As you say, it's as much about the world of the writers as it is about 'truth'. The imagery of the Communion does not trouble me because Christ in that mode is almost fused with us, the subject/object divide is no longer there in quite the same way. One possible aspect of it that is troubling from a vegan point of view is the idea of ingesting in order to take on the power of the eaten – but that idea isn't confined to those who eat animals – we think it about plant-eating too. In the Communion liturgy we hear the ostensible voice of Christ offering himself, which is not the same as a creature being killed without any say in the matter. Now people could argue that this enables us to accept the false idea that animals 'offer' themselves to us in nature! – but this would be distorting Christ's words way out of context: 'greater love has no one than to lay down their life for another', to paraphrase a translation, does not mean we should expect others to, but that we should be striving towards that ideal selflessness ourselves! 'Love your neighbour as yourself' – 'who is my neighbour?' These are very basic and resonant ideas. I don't see any incompatibility with veganism there – nor in other religions – only on the literal, legalistic plane perhaps.

But I think it's very important to stress that, while veganism may have a spiritual potential, it is *not* a religion. It's perfectly possible to approach it from a completely rationalistic, materialist viewpoint, as you say – it's a logical and effective strategy for reducing all kinds of problems in the world.

John: I think children should have the right to choice when it comes to whether or not they wish to be vegans, and most children in my experience prefer not to eat animal products, especially if they actually understand what's going on. Of course, should a child select a non-vegan diet in a vegan house, numerous other ethical questions come into play. I would never, for example, prepare a non-vegan meal. What would you do?

Tracy: I wasn't vegan when our child was born. If I'd had a child since becoming vegan, there's no doubt I would be raising him or her on a vegan diet. If the child 'rebelled' by eating other things elsewhere, I would have to think how to deal with that.... But at home nothing would change (I could not make my home in a non-vegan household. I could not cook flesh or what Carol Adams calls 'feminised protein', taken from mother cows or hens. It's just not a possible option anymore.) But I can't imagine a child realistically wanting to leave the vegan diet, except perhaps on reaching their teens, because people say that's the phase when everyone rebels against their parents! Once of an age really to make their own decisions, children should have free choice over this. As I said, I wasn't a vegan when our child was born, so I did not take then the path I might now.

Non-vegans might think this is exceptionally strict, that the child should 'have choice' from the beginning; yet few that I've met would allow their own children the same 'choice' in the opposite direction! Rather, we've met quite a few kids who say, 'I'd like to be vegetarian, but my parents won't let me'. There are two issues going on here – the moral choice, and the parenting question. Good parents of children on any diet surely make decisions *for* them, based on what they believe to be right (not too many sweets, fats, additives, etc.). I don't see how that differs with veganism, just because it happens to be a minority belief. Even if you look at it from the point of view only of protecting your child's health, I would feel irresponsible to let a child eat, say, beef, in this BSE-ridden environment – just for an example. And dairy is probably not above reproach there either. Flesh-eating parents do not doubt their right to decide for their children even after the child is 'old enough' to be vegetarian if they wish – in my experience the parents go on even then having their input (heeded or not). I think this is something we don't need to be defensive about. If they say that the child will feel 'excluded' or 'different', well there are lots of ways that can happen, and they are not unique to vegans; in fact that accompanies any life with moral fibre. You can't always resemble everyone else, and why must you? You can't always please everyone else either.

John: It is easy to condemn others for not believing what you believe yourself. I am steady in my beliefs, but am wary of fundamentalism. What is right for me might not be right for others. I've mentioned cultural difference and cultural respect. This is vital to me. How do you feel?

Tracy: Fundamentalism seems to me like a mental disease. It could infect veganism like anything else. I agree, and I do not 'preach' veganism to others in any overt sense (if my aim were to convert, I doubt that would work anyway – it's certainly not what made me a vegan). What helped me become vegan was the *example* of others. Plus reading about it, and I can see that I was prepared for that by intense immersion in 'green' ideas in the eighties, when I first tried being vegetarian.

I believe veganism is right and good, but it is not my business to judge others for what they do or don't do – I am morally responsible for my own behaviour, not theirs. And for guiding my child. No one else is my child, if you see what I mean.

I do believe, though, that flesh-eating culture is far more fundamentalist and domineering than any expression of veganism that I've encountered (although some of what I read about 'straight-edge' worries me – but I've only read about it in the mainstream press, which is likely to distort it – I've never met a straight-edger).

You know the way that heterosexual people often complain, even if only mildly, 'I don't mind gays, I just wish they wouldn't be so in-your-face, I don't want to see and hear about it all the time', and yet don't realise that gays have to see and hear about heterosexuality all the time – it is *much* more in-your-face, pervasive, loud and even punishing if you don't adhere to it. (The slightest slip and a person can be outcast – isn't that fundamentalist?) Well, I think a similar thing operates along the flesh-eating/vegan divide – we *look* visible and loud to them because they don't see their own world-view as relative, they see it as 'natural', the norm, the base.

So I do not like fundamentalism, but I am aware that my beliefs and practices will seem like it to some.

John: Linguistic disobedience is undermining the bigotry of language. We have often discussed how literature should be reread in the light of veganism – the cruelty and indifference to animals, the sheer lack of awareness as to what drives expression. In the same way that we reread in the light of the politics of ethnicity and gender, so should we via veganism.

Tracy: Yes, Carol Adams of course in *The Sexual Politics of Meat* began to do this, and refers to others who pursue it. I would like to take it even further:

go back to all the so-called greats I studied at school and see what they tell me about animal suffering and death – as well as animal life! – about human detachment and whether compassion or empathy across species is possible. It's an element that's most often ignored when a 'general' book is discussed. It would be fruitful to look at a book like *Wuthering Heights*, say, and ask what its message about cruelty begetting cruelty might be, and at what points it justifies or compromises.... Anne Brontë too links cruelty to animals with human selfishness. How does this play itself out in the imagery of consumption in the novel: where does the writer stop short of full acknowledgement of the issue? What about a book like Atwood's *The Edible Woman*, in which vegetarianism seems to be portrayed as a kind of neurotic identifying-downward that can only end in self-starvation unless we take charge and become the powerful, etc.? (Lots of Atwood circles these issues – never entirely satisfactorily, but engagingly. And, I think, with more compromise as the novels go on appearing. But this would take a lot more thinking through.)

John: People seem to respect consistency. When I was younger, my aim was to convert people – the more unlikely the person, the stronger my enthusiasm. I am embarrassed to look back on this. It's that powerlessness thing really. Which isn't to say I wasn't total in my conviction – I was! But it's easy to forget that other motives work away subtextually. Anyway, by being consistent in what I say and what I do, by being loyal and reliable, at least to the best of my ability, I find that even the most anti-vegan person develops a respect for my personal space. And that's all I can ask, is that they respect my space. A good example is better than cowardly acts such as spray-painting a wall, or silent acts of sabotage. If one must act, do so with words – and with words that don't just reinvent the genres of aggression. Revolutions end themselves before they begin – when the rhetoric comes they undermine their own integrity. And the rhetoric always comes. They are as much about control as the systems they wish to replace. Even within the State, communities can exist that operate along vegan anarchist lines. Examples that pave the way for change on a more substantial level. Suffering will not end unless there is at least a basic consensus. Force solves nothing.

References

Adams, Carol. *The Sexual Politics of Meat: A Feminist-Vegetarian Critical Theory*. London: Continuum, 1990.

Atwood, Margaret. *Power Politics*. New York: Harper & Row, 1971.

Auerbach, Eric. *Mimesis: The Representation of Reality in Western Literature*. Princeton: Princeton University Press, 2003.

Bachelard, Gaston. *Air and Dreams: An Essay on the Imagination of Movement*, trans. E.R. and C.F. Farrell. Dallas: Dallas Institute Publications, 1988.

Bakunin, Mikhail. *God and the State*. <http://www.marxists.org/reference/archive/bakunin/works/godstate/index.htm>.

Beddoes, Thomas Lovell. *Death's Jest-Book*, ed. Michael Bradshaw. Manchester: Fyfield Books, 2003.

Bookchin, Murray. *The Ecology of Freedom: The Emergence and Dissolution of Hierarchy*. New York: Black Rose Books, 1991.

Borst, Raymond R. *The Thoreau Log: A Documentary Life of Henry David Thoreau 1817–1862*. New York: Twayne, 1995.

Brady, Veronica. *Caught in the Draught: On Contemporary Australian Society and Culture*. Sydney: Angus & Robertson, 1994.

Brennan, C.J. *Poems 1913*. Sydney: Angus & Robertson, 1992.

Brewster, A., A. O'Neill and R. Van Den Berg, eds. *Those Who Remain Will Always Remember: An Anthology of Aboriginal Writing*. Fremantle: Fremantle Arts Centre Press, 2000.

Brock, Peter. *Pacifism in Europe to 1914*. Princeton: Princeton University Press, 1972.

Burgwinkle, William E., trans. *Razos and Troubadour Songs*. New York: Garland, 1990.

Cameron, Norman. *Collected Poems 1905–1953*. London: Hogarth Press, 1957.

Carson, Alfred. 'Henry Clay: A Western Australian Poet'. *Early Days: Journal of the Royal Western Australian Historical Society* 3 (1938): 61–63.

Chambers, Ross. 'Training for Modernity: Verlaine's Le Paysage dans le cadre des portières…', in *Nineteenth-Century French Poetry: Introductions to Close Reading*, ed. Christopher Prendergast. Cambridge: Cambridge University Press, 1990, pp. 157–77.

Clark, Axel. *Christopher Brennan: A Critical Biography*. Carlton: Melbourne University Press, 1990.

Cohen, David. 'Writing, Law, and Legal Practice in the Athenian Courts', in *Written Texts and the Rise of Literate Culture in Ancient Greece*, ed. Harvey Yunis. Los Angeles: University of California Press, 2003, pp. 78–96.

Coughran, Chris and Nathan Laurent. 'Please Take Me Home', in *Vagabond Holes: David McComb and The Triffids*, eds Chris Coughran and Niall Lucy. Fremantle: Fremantle Press, 2009, pp. 175–80.

Curtis, S.J. and M.E.A. Boultwood. *A Short History of Educational Ideas*. London: University Tutorial Press, 1970.

Dante. *The Divine Comedy*. <http://www.divinecomedy.org/divine_comedy.html>.

Dawe, Bruce. *Sometimes Gladness*, 3rd ed. Melbourne: Longman Cheshire, 1998.

Debord, Guy. *Society of the Spectacle*, trans. Ken Knabb. London: Rebel Press, 1983.

Defoe, Daniel. *A Journal of the Plague Year*. Oxford: Oxford University Press, 2009.

Deleuze, Gilles and Felix Guattari. *A Thousand Plateaus: Capitalism and Schizophrenia*, trans. Brian Massumi. Minneapolis: University of Minnesota Press, 1987.

Derrida, Jacques. 'The Ends of Man', in *Margins of Philosophy*, trans. Alan Bass. Chicago: University of Chicago Press, 1985, pp. 109–36.

Derrida, Jacques. 'Force of Law: The "Mystical Foundation of Authority"', in *Deconstruction and the Possibility of Justice*, eds Drucilla Cornell, Michael Rosenfeld and David Grey Carlson. London: Routledge, 1992, pp. 1–67.

Derrida, Jacques. 'Poetics and Politics of Witnessing', trans. Rachel Bowlby, in *Sovereignties in Question: The Poetics of Paul Celan*, eds Thomas Dutoit and Outi Pasanen. New York: Fordham University Press, 2005, pp. 65–96.

Dick, Philip K. *The Philip K. Dick Reader*. London: Orion, 2007.

Fagan, Kate, John Kinsella and Peter Minter, eds. *Ecopoetical Pedagogies*, special issue of *Angelaki: Journal of Theoretical Humanities* 14.2 (2009).

Freud, Sigmund. *The Uncanny*, trans. David McLintock. Harmondsworth: Penguin, 2003.

Gallie, W.B. *Philosophers of Peace and War: Kant, Clausewitz, Marx, Engels and Tolstoy*. Cambridge: Cambridge University Press, 1978.

Gelder, Ken and Jane Jacobs. *Uncanny Australia: Sacredness and Identity in a Postcolonial Nation*. Melbourne: Melbourne University Press, 1998.

Green, H.M. *A History of Australian Literature Pure and Applied: A Critical Review of All Forms of Literature Produced in Australia from the First Books Published After the Arrival of the First Fleet Until 1950*. 2 vols, revised by Dorothy Green. London: Angus & Robertson, 1984.

Grono, William, ed. *Margins: A West Coast Selection of Poetry, 1829–1988*. Fremantle: Fremantle Arts Centre Press, 1988.

Haraway, Donna. *Simian, Cyborgs and Women: The Reinvention of Nature*. New York: Routledge, 1991.

Harding, Jeremy. *The Uninvited: Refugees at the Rich Man's Gate*. London: Profile Books, 2000.

Heaney, Seamus. *North*. London: Faber, 1975.

Hewett, Dorothy. *Collected Poems*. Fremantle: Fremantle Arts Centre Press, 1995.

Hodge, Robert and Vijay Mishra. *Dark Side of the Dream: Australian Literature and the Postcolonial Mind*. Sydney: Allen & Unwin, 1991.

Hope, A.D. *Selected Poetry and Prose*, ed. David Brooks. Sydney: Halstead Press, 2000.

Hull, Coral. *The Broken Land: 5 Days in Bre, 1995*. Wollongong: Five Islands Press, 1997.

Hull, Coral. *William's Mongrels*, in Coral Hull, David Curzon, Philip Hammial and Steven Oliver, *The Wild Life*. Ringwood: Penguin, 1996.

Irigaray, Luce. *To Speak Is Never Neutral*, trans. Gail Schwab. London: Continuum, 2002.

Kac, Eduardo. 'Biopoetry'. *The Notre Dame Review* 19 (winter 2005): 145–48.

Kant, Immanuel. *Political Writings*, ed. H.S. Reiss. Cambridge: Cambridge University Press, 1991.

Kinsella, John. *Contrary Rhetoric: Lectures on Landscape and Language*, eds Glen Phillips and Andrew Taylor. Fremantle: Fremantle Press, 2008.

Kinsella, John. *Disclosed Poetics: Beyond Landscape and Lyricism*. Manchester: Manchester University Press, 2008.

Kinsella, John. *Divine Comedy: Journeys Through a Regional Geography*. New York: W.W. Norton, 2008.

Kinsella, John, ed. *Dreamhoard: Pipe Dreams, Daydreams, Reveries and Nightmares*. Cambridge: Salt Publishing, forthcoming.

Kinsella, John. *Full Fathom Five*. Fremantle: Fremantle Arts Centre Press, 1993.

Kinsella, John. 'Geography Is History?' *Poetry Review* 60.2 (2000): 4–6.

Kinsella, John. *Icononotes*. (Icons by Alexander Deriev.) Stockholm: Ars Interpres, forthcoming.

Kinsella, John. *Love Sonnets*. Cambridge: Equipage, 2005.

Kinsella, John. *The New Arcadia: Poems*. New York: W.W. Norton, 2005.

Kinsella, John. *Night Parrots*. Fremantle: Fremantle Arts Centre Press, 1989.

Kinsella, John, ed. *The Penguin Anthology of Australian Poetry*. Sydney: Penguin, 2009.

Kinsella, John. *Poems 1980–1994*. London: Bloodaxe Books, 1998.

Kinsella, John. *The Silo: A Pastoral Symphony*. Fremantle: Fremantle Arts Centre Press, 1995.

Kinsella, John. *Visitants*. London: Bloodaxe Books, 1999.

Kinsella, John and Coral Hull. *Zoo*. Sydney: Paper Bark, 2000.

Kinsella, John and Niall Lucy, eds. *Plagiarism! (From Work to* Détournement*)*, special issue of *Angelaki: Journal of Theoretical Humanities* 14:1 (2009).

Kinsella, John and Ron Sims. *Kangaroo Virus* (includes CD). Fremantle: Fremantle Arts Centre Press, 1998.

Kropotkin, Peter. *Mutual Aid: A Factor of Evolution*. <http://dwardmac.pitzer.edu/Anarchist_archives/kropotkin/mutaidcontents.html>.

Lea, Bronwyn. *The Other Way Out: New Poems*. Artarmon: Giramondo, 2008.

Lean, Geoffrey and Harriet Shawcross. 'Are Mobile Phones Wiping Out Our Bees?' *Independent*, 15 April 2007. <http://www.independent.co.uk/environment/nature/are-mobile-phones-wiping-out-our-bees-444768.html>.

Little, Lorna. *The Mark of the Wagarl*. Illustrated by Janice Lyndon. Broome: Magabala Books, 2004.

Lucy, Niall. *A Derrida Dictionary*. Oxford: Blackwell, 2004.

Lucy, Niall. 'Running On/BORDER LINES'. *Derrida Today* 1.2 (2008): 236–53.

Malatesta, Errico. *Anarchy*. <http://www.marxists.org/archive/malatesta/1891/xx/anarchy.htm>.

Malatesta, Errico. *Majorities and Minorities*. <http://www.marxists.org/archive/malatesta/unk/xx/majmin.htm>.

Malatesta, Errico. *Revolution in Practice*. <http://dwardmac.pitzer.edu/anarchist_Archives/malatesta/revpra.html>.

Manne, Robert, ed. *Whitewash: On Keith Windschuttle's Fabrication of Aboriginal History*. Melbourne: Black Inc., 2003.

Marr, David and Marian Wilkinson. *Dark Victory*. Sydney: Allen & Unwin, 2003.

Marti, Felix. 'Understanding and Dialogue Between Religions to Promote the Spirit of Peace', in *From A Culture of Violence to A Culture of Peace: Peace and Conflict Issues*, ed. collective. Paris: UNESCO, 1996, pp. 145–61.

McMillan, Margaret. *Education Through the Imagination*. London: S. Sonnenschein, 1904.

Miller, E. Morris. *Australian Literature: A Bibliography to 1938*. Extended to 1950, ed. Frederick T. Macartney. Sydney: Allen & Unwin, 1956.

Motte, Warren F. Jr. *Questioning Edmond Jabès*. Lincoln: University of Nebraska Press, 1990.

Murray, Les. *The People's Otherworld*. Sydney: Angus & Robertson, 1983.

O'Reilly, John Boyle. *Complete Poems and Speeches*. New York: Cassell, 1891.

Page, Geoff. *Seriatim*. Cambridge: Salt Publishing, 2007.

Propp, Vladimir. *Morphology of the Folktale*, trans. Laurence Scott. Austin: University of Texas Press, 1968.

Roseneil, Sasha. *Disarming Patriarchy: Feminism and Political Action at Greenham*. Buckingham: Open University Press, 1995.

Scigaj, Leonard M. *Sustainable Poetry: Four American Ecopoets*. Lexington: Kentucky University Press, 1999.

Sidney, Philip. *The Countess of Pembroke's Aracadia (The Old Arcadia)*. Oxford: Oxford University Press, 1999.

Slovic, Paul. *The Perception of Risk*. London: Earthscan, 2000.

Stendhal. *Lucien Leuwen*, trans. H.L.R. Edwards. Harmondsworth: Penguin, 1991.

Thomas, R.S. *Collected Poems 1945–1990*. London: Phoenix, 2000.

Thoreau, Henry David. *Walden*. <http://www.gutenberg.org/etext/205>.

Todorov, Tzvetan. *The Fantastic: A Structural Approach to a Literary Genre*, trans. Richard Howard. Ithaca: Cornell University Press, 1975.

Verlaine, Paul. *La bonne chanson*. Paris: Creuzvault, 1936.

Ward, Colin and David Goodway. *Talking Anarchy*. Nottingham: Five Leaves, 2003.

Wark, McKenzie. *A Hacker Manifesto*. Cambridge: Harvard University Press, 2004.

Wark, McKenzie. *Virtual Geography: Living with Global Media Events*. Bloomington: Indiana University Press, 1994.

Windschuttle, Keith. *The Fabrication of Aboriginal History, Volume One: Van Diemen's Land 1803–1847*. Sydney: Macleay Press, 2002.

Wright, Judith. *Preoccupations in Australian Poetry*. Melbourne: Oxford University Press, 1966.

Zahavi, Helen. *Dirty Weekend*. Minneapolis: Cleis Press, 2000.

Index

www.ingramcontent.com/pod-product-compliance
Lightning Source LLC
LaVergne TN
LVHW022204200926
848481LV00006B/1839